Communications in Computer and Information Science

673

Commenced Publication in 2007
Founding and Former Series Editors:
Alfredo Cuzzocrea, Dominik Ślęzak, and Xiaokang Yang

More information about this series at http://www.springer.com/series/7899

Viktor V. Krasnoproshin
Sergey V. Ablameyko (Eds.)

Pattern Recognition and Information Processing

13th International Conference, PRIP 2016
Minsk, Belarus, October 3–5, 2016
Revised Selected Papers

 Springer

Editors
Viktor V. Krasnoproshin
Belarusian State University
Minsk
Belarus

Sergey V. Ablameyko
Belarusian State University
Minsk
Belarus

ISSN 1865-0929   ISSN 1865-0937   (electronic)
Communications in Computer and Information Science
ISBN 978-3-319-54219-5   ISBN 978-3-319-54220-1   (eBook)
DOI 10.1007/978-3-319-54220-1

Library of Congress Control Number: 2017932121

Printed on acid-free paper

This Springer imprint is published by Springer Nature
The registered company is Springer International Publishing AG
The registered company address is: Gewerbestrasse 11, 6330 Cham, Switzerland

Preface

The International Conference on Pattern Recognition and Information Processing (PRIP) started its history in 1991 in Minsk.

Belarusian research in the area of pattern recognition and image processing was very intensive and productive in the former Soviet Union. As a result, the USSR Association of Pattern Recognition decided to organize the First All-Union Conference on "Pattern Recognition and Image Analysis." The proposal to host the conference was made to Belarusian scientists. The conference was held in October 1991 in Minsk, with more than 200 researchers participating.

In December 1992, after the collapse of the USSR, the Belarusian Association for Image Analysis and Recognition (BAIAR) was founded and in March 1993 the International Association of Pattern Recognition (IAPR) officially accepted BAIAR as a national representative of Belarus.

BAIAR management decided that the second conference in the field of pattern recognition and image processing should take place. BAIAR formulated the goal of the PRIP conference: "To establish cooperation between Belarusian researchers and [the] international community in the field of pattern recognition and image analysis." Since then, the conference has been held every two consecutive years. The conference in 2016 was the 13th event in this series.

Today, PRIP conferences are well known and well recognized. Information about PRIP is included on all major websites on computer vision and pattern recognition. Conference proceedings are cited in the INSPEC, the main global database of publications.

In 2001, BAIAR decided that English would be the only conference language.

PRIP is held in cooperation with other scientific establishments. PRIP 2016, like previous conferences, was endorsed by the IAPR.

In all, 72 papers were submitted for PRIP 2016 from 15 countries (132 authors). All submissions were reviewed by Program Committee members together with two additional reviewers. As a result, 49 papers were selected for inclusion in the PRIP 2016 scientific program.

PRIP 2016 had two categories of sessions: plenary and regular sessions. At plenary sessions, presentations were made by six inviter speakers, including the president of IAPR Ingela Nystrom (Sweden). All the sessions were held in a single track and participants had enough time for heated discussions after each presentation.

The proceedings of PRIP conferences are published regularly by the conference organizers. This year, a collection of selected papers (18 papers), among those accepted for the program of the PRIP conference, after final reviewing were recommended for publishing in Springer's *Communications in Computer and Information Science* (CCIS) series.

PRIP 2016 was held at the Belarusian State University, one of the leading and oldest Belarusian universities, during October 3–5, 2016. The conference was dedicated to the 60th birthday of the academician and rector of the Belarusian State University Sergey Ablameyko.

The PRIP 2016 proceedings present new results in the area of pattern recognition and image processing and its applications.

The volume is aimed at researchers working in pattern recognition and image analysis, knowledge processing, and knowledge-based decision support system.

January 2017

Sergey Ablameyko
Victor Krasnoproshin

Organization

The International Conference on Pattern Recognition and Information Processing (PRIP) was hosted by the Belarusian State University, Faculty of Applied Mathematics and Computer Sciences, in cooperation with the Belarusian State University of Informatics and Radioelectronics, the United Institute of Informatics Problems of the National Academy of Sciences of Belarus, and the Belarusian Association for Image Analysis and Recognition.

Conference Chairs

S. Ablameyko   Belarus
V. Krasnoproshin   Belarus

Conference Co-chairs

A. Tuzikov   Belarus
D. Samal   Belarus

Program Committee

G. Borgefors   Uppsala University, Sweden
K. Boyer   College of Engineering and Applied Sciences, USA
S. Demidenko   Massey University, New Zealand
V. Donskoy   Crimean Federal University, Russia
A. Doudkin   National Academy of Sciences, Belarus
M. Frucci   Institute of High Performance Computing
                  and Networking, Italy
J. Gil Aluha   Royal Academy of Economic and Financial Sciences, Spain
V. Golenkov   Belarusian State University of Informatics
                  and Radioelectronics, Belarus
V. Golovko   Brest State Technical University, Belarus
I. Gurevich   Russian Academy of Sciences, Russia
R. Hiromoto   University of Idaho, USA
A. Imada   Brest State Technical University, Belarus
Y. Kharin   Belarus State University, Belarus
Y. Kondratenko   Black Sea State University, Ukraine
K. Madani   University of Paris-Est Créteil Val de Marne, France
A. Marcelli   University of Salerno, Italy
I. Nystrom   Uppsala University, Sweden
T. Pridmore   University of Nottingham, UK
V. Ryazanov   Moscow Institute for Physics and Technology, Russia

G. Sanniti di Baja	Institute of High Performance Computing and Networking, Italy
A. Sashenko	Ternopil National Economic University, Ukraine
G. Shakach	Ajloun National University, Jordan
V. Shmerko	University of Calgary, Canada
I. Sovpel	Belarus State University, Belarus
S. Uchida	University of Tokyo, Japan
E. Zaitseva	University of Žilina, Slovakia
Y. Zhuravlev	Moscow State University, Russia

Additional Reviewers

J. Gil Lafuente, Spain
J. Zuev, Russia
T. Downs, Australia
A. Kurbatski, Belarus
S. Janushkevich, Canada
N. Tung, Vietnam
D. Samal, Belarus
H. Liu, China
L. Aslanyan, Armenia
B. Fioleu, France

Local Organizing Committee

V. Obraztsov - Chair
V. Konakh
T. Korbut
N. Roubashko
A. Valvachov

In collaboration with:

- FE Byelex Multimedia Products Ltd.
- International Associations for Pattern Recognition

Contents

Information Processing and Applications

Summarizing Lectures (Plenary Papers)

BoneSplit - A 3D Painting Tool for Interactive Bone Segmentation in CT Images

Ingela Nyström[1(✉)], Johan Nysjö[1], Andreas Thor[2], and Filip Malmberg[1]

[1] Centre for Image Analysis, Uppsala University, Uppsala, Sweden
ingela.nystrom@it.uu.se
[2] Department of Surgical Sciences, Uppsala University Hospital, Uppsala, Sweden

Abstract. We present an efficient interactive tool for segmenting individual bones and bone fragments in 3D computed tomography (CT) images. The tool, which is primarily intended for virtual cranio-maxillofacial (CMF) surgery planning, combines direct volume rendering with interactive 3D texture painting to enable quick identification and marking of bone structures. The user can paint markers (seeds) directly on the rendered bone surfaces as well as on individual CT slices. Separation of the marked bones is then achieved through the random walks algorithm, which is applied on a graph constructed from the thresholded bones. The segmentation runs on the GPU and can achieve close to real-time update rates for volumes as large as $512 \times 512 \times 512$ voxels. The user can perform segmentation editing to correct the result. An evaluation reports segmentation results comparable with manual segmentations, but obtained within a few minutes. In the invited PRIP talk, BoneSplit is presented and how the tool fits into our haptics-assisted surgery-planning system.

1 Introduction

Cranio-maxillofacial (CMF) surgery to restore the facial skeleton after serious trauma or disease can be both complex and time-consuming. There is, however, evidence that careful virtual surgery planning can improve the outcome and facilitate the restoration [18]. In addition, virtual surgery planning can lead to reduced operating times and costs.

Recently, a system for planning the restoration of skeletal anatomy in facial trauma patients has been developed within our research group [17]. As input, the system requires segmented 3D computed tomography (CT) data from the fractured regions, in which individual bone fragments are labeled (Fig. 2c). While a collective bone segmentation can be obtained relatively straightforward by, for instance, thresholding the CT image at a Hounsfield unit (HU) value corresponding to bone tissue, segmentation of individual bone structures is a more difficult and time-consuming task. Due to bone density variations and image imprecisions such as noise and partial volume effects, adjacent bones and bone fragments in CT images are often connected to each other after thresholding and cannot be separated by simple connected component analysis or morphological operations.

© Springer International Publishing AG 2017
V.V. Krasnoproshin and S.V. Ablameyko (Eds.): PRIP 2016, CCIS 673, pp. 3–13, 2017.
DOI: 10.1007/978-3-319-54220-1_1

In the current planning procedure, the bones are separated manually, slice by slice, using the brush tool in the ITK-SNAP software [19]. This process takes several hours to complete and is a major bottleneck in the planning.

1.1 Contribution

In this paper, we present BoneSplit [16], an efficient interactive tool for segmenting bones and bone fragments in CT images. The tool combines direct volume rendering with an interactive 3D painting interface to enable quick identification and marking of bone structures. The user can paint markers (seeds) directly on the rendered bone surfaces as well as on individual CT slices. Separation of the marked bones is then achieved with a graph-based segmentation algorithm [9] and a local 3D editing tool.

1.2 Related Work

Automatic bone segmentation methods with or without shape priors have been proposed [2,12] and used for segmentation of individual intact bones such as the tibia or fibula, but are not general or robust enough for fracture segmentation. Manual segmentation, on the other hand, can produce accurate delineations of fractured bones and is often used in CMF planning studies, but is too tedious and time-consuming to perform for routine clinical usage and suffers from low repeatability. Another problem with manual segmentation is that the user only operates on a single slice at a time and does not perceive the full 3D structure of the bones, which often results in irregular segmentation boundaries.

Semi-automatic or interactive segmentation methods aim to overcome the limitations of automatic and manual segmentation by combining high-level user input with efficient and exact delineation algorithms. This approach can often yield accurate and repeatable segmentation results while reducing the amount of manual interaction. An example of a general-purpose interactive segmentation tool is GeoS [5]. Liu et al. [15] used a graph cut-based [3] technique to separate collectively segmented bones in the foot, achieving an average segmentation time of 18 min compared with 1.5–3 h for manual segmentation. Fornaro et al. [7] and Fürnstahl et al. [8] combined graph cuts with a bone sheetness measure [6] to segment fractured pelvic and humerus bones, respectively.

In current semi-automatic segmentation tools, the user typically interacts with the segmentation via 2D slice views. A problem with slice-based interaction, however, is that it can be difficult to locate and mark individual bone structures, particularly in complex fracture cases. Texture painting tools [11] enable efficient and intuitive painting of graphical models (3D meshes) by mapping 2D mouse strokes in screen space to 3D brush strokes in object space. Mesh segmentation methods [14] utilize similar sketch-based interfaces for semi-automatic labeling of individual parts in 3D meshes. Bürger et al. [4] developed a direct volume editing tool that can be used for manual labeling of bone surfaces in CT images. Our work extends the 3D painting concept to interactive bone segmentation.

2   Methods

Our segmentation tool combines and modifies several image analysis and visualization methods, which are described in the following sections. In brief, the main steps are (1) collective bone segmentation, (2) marking of individual bone structures, (3) random walks bone separation, and (4) segmentation editing.

2.1   Collective Bone Segmentation

A collective bone segmentation is obtained by thresholding the grayscale image at the intensity value t_{bone} (see Fig. 1). The threshold is preset to 300 HU in the system, but can be adjusted interactively, if needed, to compensate for variations in bone density or image quality. The preset value was determined empirically and corresponds to the lower HU limit for trabecular (spongy) bone. Noisy images can be smoothed with a $3 \times 3 \times 3$ Gaussian filter ($\sigma = 0.6$) prior to thresholding. The Gaussian filter takes voxel anisotropy into account and can be applied multiple times to increase the amount of smoothing.

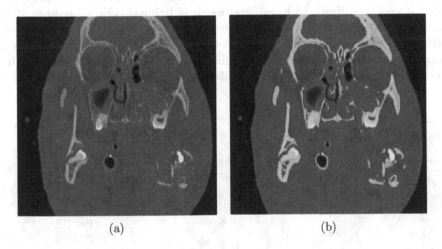

(a)   (b)

Fig. 1. Left: Coronal slice of a 3D CT image of the facial skeleton in a complex trauma case. Right: Collective bone segmentation obtained by thresholding the CT image at a Hounsfield unit (HU) value corresponding to trabecular bone.

2.2   Deferred Isosurface Shading

We use GPU-accelerated ray-casting [13] to render the bones as shaded isosurfaces. The isovalue is set to t_{bone}, so that the visual representation of the bones matches the thresholding segmentation. Similar to [10], we implement a deferred isosurface shading pipeline (see [16] for more details.) A 32^3 min-max block volume is used for empty-space skipping and rendering of the ray-start positions. We render the first-hit positions and surface normals to a G-buffer via multiple render targets (MRT), and calculate shadows and local illumination in additional passes. Segmentation labels are stored in a separate 3D texture and fetched with nearest-neighbor sampling in the local illumination pass.

Local illumination is calculated using a normalized version of the Blinn-Phong shading model [1]. To make it easier for the user to perceive depth and spatial relationships between bones and bone fragments, we combine the local illumination with shadow mapping to render cast shadows. The shadow map is derived from an additional first-hit texture rendered from a single directional light source's point of view.

Ambient lighting is provided from pre-filtered irradiance and radiance cube maps [1]. The color and intensity variations in the image-based ambient lighting allow the user to see the shape and curvature of bone structures that are in shadow. The ambient lighting is combined with ambient occlusion [1] (self-shadowing) to visually enhance fracture locations.

2.3 3D Texture Painting Interface

As stated in Sect. 1.2, a problem with conventional 2D slice-based interaction is that it can be difficult to locate and mark individual bone structures. Even radiologists, who are highly skilled in deriving anatomical 3D structures from stacks of 2D images, may find it difficult to locate and mark bone fragments in complex fracture cases. To overcome this issue, we implemented a 3D texture painting interface that allows the user to draw seeds directly on the bone surfaces.

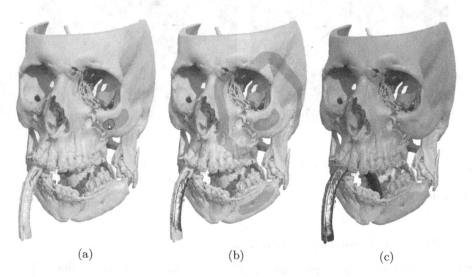

(a)   (b)   (c)

Fig. 2. Interactive bone separation: (a) 3D brush used for painting seeds directly on the rendered bone surfaces; (b) marked bones; (c) segmentation result obtained with the random walks algorithm.

Our 3D brush (Fig. 2a) is implemented as a spherical billboard and uses the first-hit texture from the G-buffer for picking and seed projection. The brush proxy follows the bone surface and can only paint seeds on surface regions that are visible and within the brush radius (in camera space).

Additional tools include a label picker, an eraser, a floodfill tool, and a local editing tool. A 3D slice viewer enables the user to mark occluded bones or place additional seeds inside the bones. We also provide interactive clipping tools that can be used to expose bones and contact surfaces.

2.4   Random Walks Bone Separation

Given the collective binary bone segmentation, the next step is to separate the individual bones and bone fragments. We considered two graph-based segmentation algorithms for this task: graph cuts [3] and random walks [9]. In the end, we selected the random walks algorithm since it is robust to noise and weak boundaries, extends easily to multi-label (K-way) segmentation, and does not suffer from the small-cut problem of graph cuts. The main drawback and limitation of random walks is its high computational and memory cost. For interactive multi-label segmentation of volume images, this has traditionally limited the maximum volume size to around 256^3, which is smaller than the CT volumes normally encountered in CMF planning. Our implementation overcomes this limitation by only operating on bone voxels.

We construct a weighted graph $G = (V, E)$ from the collective bone segmentation and use the random walks algorithm to separate individual bones marked by the user. Figure 2 illustrates the segmentation process. For every bone voxel, the random walks algorithm calculates the probability that a random walker starting at the voxel will reach a particular seed label. A crisp segmentation is obtained by, for each bone voxel, selecting the label with the highest probability value. The vertices $v \in V$ in the graph represent the bone voxels and the edges $e \in E$ represent the connections between adjacent bone voxels in a 6-connected neighborhood. The number of neighbors can vary from zero to six. Each edge e_{ij} between two neighbor vertices v_i and v_j is assigned a gradient magnitude-based weight w_{ij} [9] defined as

$$w_{ij} = \exp(-\beta(g_i - g_j)^2) + \epsilon, \tag{1}$$

where g_i and g_j are the intensities of v_i and v_j in the underlying grayscale image, and β is a parameter that determines the influence of the gradient magnitude. We add a small positive constant ϵ (set to 0.01 in our implementation) to ensure that v_i and v_j are connected, i.e., $w_{ij} > 0$. Increasing the value of β makes the random walkers less prone to traverse edges with high gradient magnitude. Empirically, we have found $\beta = 3000$ to work well for bone separation; however, the exact choice of β is not critical and we have used values in the range 2000–4000 with similar results.

We represent the weighted graph and the seed nodes as a sparse linear system and use a GPU-accelerated conjugate gradient (CG) solver to compute the solution for each label. Implementation details and benchmarks of the solver are provided in [16]. By constructing the graph from the bone voxels in the collective segmentation, rather than from the full image, we simplify the segmentation problem from separation of multiple tissue types to bone separation. Moreover, we reduce the memory and computational cost substantially. The head CT volumes encountered in CMF planning typically contain between 3 and 8 million bone voxels, which is a small fraction, around 10%, of the total number of voxels. Combined with fast iterative solvers, this enables rapid update of the segmentation for volumes as large as 512^3.

2.5 Segmentation Editing

The user can edit the initial random walks segmentation by painting additional seeds on the bone surfaces or CT slices and running the iterative solver again, using the previous solution as starting guess [9] to speed up convergence. This procedure can be repeated until all bone structures have been separated. Visual inspection of the result during editing is supported by volume clipping (Fig. 3).

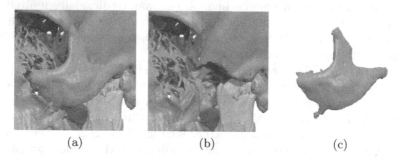

(a)   (b)   (c)

Fig. 3. To support visual inspection and editing of bone fragments and contact surfaces, a segmented region (a) can be hidden (b) or exposed (c) via volume clipping. The clipping is performed by temporarily setting the grayscale value of the segmented region to $t_{bone} - 1$ and updating the grayscale 3D texture.

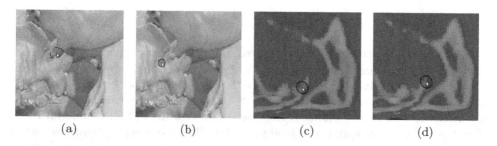

(a)   (b)   (c)   (d)

Fig. 4. Segmentation editing performed with the local 3D editing tool.

Further fine-scale editing can be performed with a dedicated 3D editing tool, which updates a local region of the segmentation in real-time and allows a selected label to grow and compete with other labels. See Fig. 4 for examples.

2.6 Implementation Details

We implemented the segmentation system in Python, using OpenGL and GLSL for the rendering, PySide for the graphical user interface, and Cython and PyOpenCL for the image and graph processing.

3 Case Study

Two users performed interactive segmentations of the facial skeleton in CT scans of three complex CMF cases. The first user was a novice on the system and received a 15 min training session, whereas the second user (the second author) was highly experienced in using the system. Both users had prior experience of manual bone segmentation but no surgical training. The CT scans were obtained as anonymized DICOM files. Further details about the datasets are provided in Table 1. Figure 5a–c show the collective bone segmentations obtained by thresholding. Bone separation was carried out in three stages:

Table 1. Details about the CT images used in the case study. The CT scans were provided by the Department of Surgical Sciences at Uppsala University Hospital.

Case	Region	Description	#Labels	Dimensions	Threshold	#Bone voxels
1	Head	Multiple fractures	15	$512 \times 512 \times 337$	260	4426530
2	Head	Multiple fractures	12	$512 \times 512 \times 301$	300	4769742
3	Head	Tumor	6	$230 \times 512 \times 512$	300	2787469

1. Initial random walks segmentation of marked bones.
2. Interactive coarse editing of the segmentation result by running random walks multiple times with additional seed strokes as input.
3. Fine-scale editing with the local 3D editing tool.

We measured the interaction time required for each stage and asked the users to save the segmentation result obtained in each stage. Additionally, one of the users segmented case 1 manually in the ITK-SNAP [19] software to generate a reference segmentation for accuracy assessment. The manual segmentation took ~5 h to perform and was inspected and validated by a CMF surgeon.

To assess segmentation accuracy and precision, we computed the Dice similarity coefficient

$$DSC = \frac{2|A \cap B|}{|A| + |B|}. \tag{2}$$

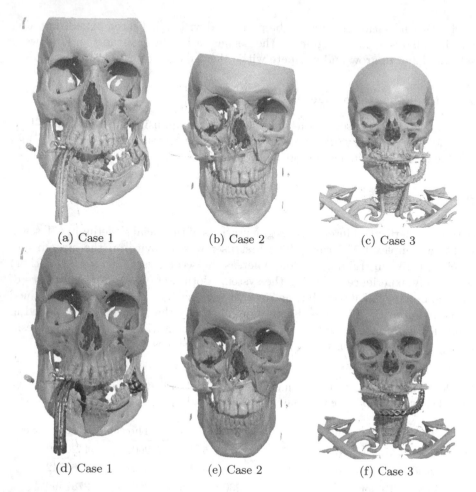

(a) Case 1   (b) Case 2   (c) Case 3

(d) Case 1   (e) Case 2   (f) Case 3

Fig. 5. Top row: Collective bone segmentations obtained by thresholding. Bottom row: Separated bones.

DSC measures the spatial overlap between two multi-label segmentations A and B and has the range $[0, 1]$, where 0 represents no overlap and 1 represents complete overlap.

The interactive segmentations (Fig. 5d–f) took on average 14 min to perform. As shown in Fig. 6, most of the time was spent in the local editing stage (stage 3). DSC between the final interactive case 1 segmentations and the manual reference segmentation was 0.97782 (User 1) and 0.97784 (User 2), indicating overall high spatial overlap. The inter-user precision (Table 2) was also high and improved with editing.

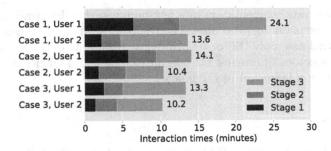

Fig. 6. Interaction times (in minutes) for the two users.

Table 2. Inter-user precision for the interactive segmentations.

Case	DSC		
	Stage 1	Stage 2	Stage 3
1	0.9199	0.9955	0.9971
2	0.9533	0.9968	0.9971
3	0.9832	0.99	0.9915

4 Discussion

Overall, the segmentation tool was found to be fast and accurate enough for the intended application. The 3D painting interface makes it easy for users to mark bone structures of interest and edit the segmentation result, and the local editing tool is useful for cleaning up the random walks segmentation and refining contact surfaces between separated bones. Two minor issues with the local editing tool were that it sometimes expanded the active label region too far or produced isolated voxels. To address the latter problem, we have recently extended the system with a connected component analysis filter that allows the user to remove small isolated components before exporting the final segmentation. Another potential problem with our current system is that the initial thresholding segmentation tends to exclude some of the thin or low-density bone structures or include noise and soft tissue. With minor modifications, the system can, however, be extended to take arbitrary binary bone segmentations as input and is thus not limited to thresholding. We also plan to extend the system with additional editing tools that enable the user to fill in holes or missing bone regions.

5 Conclusion

In this paper, we have presented an efficient 3D texture painting tool for segmenting individual bone structures in 3D CT images. This type of segmentation is crucial for virtual CMF surgery planning [17], and can take several hours to perform with conventional manual segmentation approaches. Our tool can produce an accurate segmentation in a few minutes, thereby removing a major bottleneck in the planning procedure. The resulting segmentation can be used as input for virtual assembly [17] or be printed as plastic models on a 3D printer. Future work includes improving the overall usability and efficiency of the tool and testing it on a larger set of CT images.

References

1. Akenine-Möller, T., Haines, E., Hoffman, N.: Real-Time Rendering, 3rd edn. A. K. Peters Ltd., Natick (2008)
2. Alathari, T., Nixon, M., Bah, M.: Femur bone segmentation using a pressure analogy. In: 22nd International Conference on Pattern Recognition (ICPR), pp. 972–977 (2014)
3. Boykov, Y.Y., Jolly, M.-P.: Interactive graph cuts for optimal boundary and region segmentation of objects in ND images. In: 8th IEEE International Conference on Computer Vision (ICCV), vol. 1, pp. 105–112 (2001)
4. Bürger, K., Krüger, J., Westermann, R.: Direct volume editing. IEEE Trans. Vis. Comput. Graph. **14**(6), 1388–1395 (2008)
5. Criminisi, A., Sharp, T., Blake, A.: GeoS: Geodesic image segmentation. In: Forsyth, D., Torr, P., Zisserman, A. (eds.) ECCV 2008. LNCS, vol. 5302, pp. 99–112. Springer, Heidelberg (2008). doi:10.1007/978-3-540-88682-2_9
6. Descoteaux, M., Audette, M., Chinzei, K., Siddiqi, K.: Bone enhancement filtering: application to sinus bone segmentation and simulation of pituitary surgery. Comput. Aided Surg. **11**(5), 247–255 (2006)
7. Fornaro, J., Székely, G., Harders, M.: Semi-automatic segmentation of fractured pelvic bones for surgical planning. In: Bello, F., Cotin, S. (eds.) ISBMS 2010. LNCS, vol. 5958, pp. 82–89. Springer, Heidelberg (2010). doi:10.1007/978-3-642-11615-5_9
8. Fürnstahl, P., Székely, G., Gerber, C., Hodler, J., Snedeker, J.G., Harders, M.: Computer assisted reconstruction of complex proximal humerus fractures for preoperative planning. Med. Image Anal. **16**(3), 704–720 (2012)
9. Grady, L.: Random walks for image segmentation. IEEE Trans. Pattern Anal. Mach. Intell. **28**(11), 1768–1783 (2006)
10. Hadwiger, M., Sigg, C., Scharsach, H., Bühler, K., Gross, M.: Real-time ray-casting and advanced shading of discrete isosurfaces. Comput. Graph. Forum **24**(3), 303–312 (2005)
11. Hanrahan, P., Haeberli, P.: Direct WYSIWYG painting and texturing on 3D shapes. ACM SIGGRAPH Comput. Graph. **24**(4), 215–223 (1990)
12. Krcah, M., Székely, G., Blanc, R.: Fully automatic and fast segmentation of the femur bone from 3D-CT images with no shape prior. In: IEEE International Symposium on Biomedical Imaging, pp. 2087–2090. IEEE (2011)
13. Krüger, J., Westermann, R.: Acceleration techniques for GPU-based volume rendering. In: 14th IEEE Visualization (VIS 2003), pp. 287–292 (2003)
14. Lai, Y.-K., Hu, S.-M., Martin, R.R., Rosin, P.L.: Rapid and effective segmentation of 3D models using random walks. Comput. Aided Geom. Des. **26**(6), 665–679 (2009)

15. Liu, L., Raber, D., Nopachai, D., Commean, P., Sinacore, D., Prior, F., Pless, R., Ju, T.: Interactive separation of segmented bones in CT volumes using graph cut. In: Metaxas, D., Axel, L., Fichtinger, G., Székely, G. (eds.) MICCAI 2008. LNCS, vol. 5241, pp. 296–304. Springer, Heidelberg (2008). doi:10.1007/978-3-540-85988-8_36

16. Nysjö, J., Malmberg, F., Sintorn, I.-M., Nyström, I.: BoneSplit - a 3D texture painting tool for interactive bone separation in CT images. J. WSCG **23**(2), 157–166 (2015)

17. Olsson, P., Nysjö, F., Hirsch, J.-M., Carlbom, I.B.: A haptics-assisted cranio-maxillofacial surgery planning system for restoring skeletal anatomy in complex trauma cases. Int. J. Comput. Assist. Radiol. Surg. **8**(6), 887–894 (2013)

18. Roser, S.M., et al.: The accuracy of virtual surgical planning in free fibula mandibular reconstruction: comparison of planned and final results. J. Oral Maxillofac. Surg. **68**(11), 2824–2832 (2010)

19. Yushkevich, P.A., Piven, J., Cody, H., Hazlett, H.C., Smith, R.G., Ho, S., Gee, J.C., Gerig, G.: User-guided 3D active contour segmentation of anatomical structures: significantly improved efficiency and reliability. Neuroimage **31**(3), 1116–1128 (2006)

Computer-Based Technologies for Virtual Screening and Analysis of Chemical Compounds Promising for Anti-HIV-1 Drug Design

A.M. Andrianov[1], I.A. Kashyn[1,2], and A.V. Tuzikov[2(✉)]

[1] Institute of Bioorganic Chemistry NASB, 5/2 Academician Kuprevich street,
220141 Minsk, Belarus
andrianov@iboch.bas-net.by
[2] United Institute of Informatics Problems NASB, 6, Surganov street,
220012 Minsk, Belarus
tuzikov@newman.bas-net.by

Abstract. Computer-based technologies for *in silico* drug development comprising virtual screening, high-throughput docking, molecular dynamics simulations, and binding free energy calculations are presented. The efficiency of these technologies is demonstrated by the identification of novel potential anti-HIV-1 agents able to mimic pharmacophoric properties of potent and broad neutralizing antibodies 10e8, VRC01, and 3074 that target three different functionally conserved regions of the viral envelope proteins.

Keywords: Virtual screening · Molecular docking · Molecular dynamics · Binding free energy calculations · HIV-1 entry inhibitors · Broadly neutralizing antibodies

1 Introduction

To date, over twenty five drugs have been approved by the United States Food and Drug Administration for the treatment of HIV infection (reviewed in [1, 2]). These drugs are distributed into six major classes: (1) nucleoside-analog reverse transcriptase inhibitors, (2) non-nucleoside reverse transcriptase inhibitors, (3) protease inhibitors, (4) fusion inhibitors, (5) entry inhibitors, and (6) integrase inhibitors [1, 2]. The majority of these anti-HIV drugs belong to the inhibitors of reverse transcriptase and protease [1, 2]. These inhibitors act inside a target cell and cannot block the initial steps of the HIV-1 life cycle associated with virus entry. In this context, development of novel, potent and broad-spectrum HIV-1 entry inhibitors is an area of considerable interest in the current anti-HIV drug design and discovery.

HIV-1 infection begins with virion entry into target cells through the interaction of viral envelope (Env) protein gp120 with primary receptor CD4 (reviewed in [3]). The binding of gp120 to CD4 induces the exposure of a second binding site for cellular co-receptor CCR5 or CXCR4 [3]. Following the binding, the gp41 transmembrane subunit of the Env protein undergoes a dramatic conformational change to mediate virus-cell

V.V. Krasnoproshin and S.V. Ablameyko (Eds.): PRIP 2016, CCIS 673, pp. 14–23, 2017.
DOI: 10.1007/978-3-319-54220-1_2

membrane fusion, enabling the virus capsid to enter the cell [3]. Many small molecule inhibitors that block the virus adsorption onto the host cell membrane and/or cell-mediated fusion have been developed [4]. However, the majority of these inhibitors have failed to be useful in clinical practice. Despite these disappointing results, the design of (+)-DMJ-I-228 and (+)-DMJ-II-121 inhibitors that target the Env trimer and present functional antagonists of viral entry [5] gave hope of future success in the development of novel efficient anti-HIV-1 drugs. This hope was supported by the discovery of anti-HIV-1 broadly neutralizing antibodies (bNAbs) and their specific modes of recognition on the viral Env, providing a new strategy for improved vaccine and drug design (reviewed in [6–8]). Neutralizing antibodies tend to increase in potency over time and broadly cross neutralizing responses, capable of recognizing heterologous HIV-1 variants, develop in a subset of individuals after primary infection. In some cases, the specificities of the antibodies conferring breadth have been mapped and are reactive with conserved Env regions.

In light of discovering anti-HIV-1 bNAbs [6–8], studies aimed at the identification of small molecules able to mimic pharmacophoric properties of these antibodies are of great challenge. In doing so, the latest computational technologies for structure-based drug design can be used at the first steps of solving this problem to significantly reduce drug development time and financial costs.

Recently, computer-aided methods of three-dimensional (3D) structure modeling and the study of the quantitative relationship between structure and biological activity of chemical substances (QSAR; Quantitative Structure-Activity Relationships) have occupied an important place in anti-HIV-1 drug development (reviewed in [9]). One of the more promising methods of computational drug design is molecular databases virtual screening aimed at the discovery of active structures. This method includes the following stages: (i) a choice of biological target and modeling of its high-resolution 3D structure; (ii) selection of structural databases of organic substances and the search for specific compounds which, according to the information on their structural and physicochemical properties are capable of exhibiting the biological activity to the prescribed target; (iii) molecular docking of the selected substances with the biological target using software based on criterion functions and choice of the potential ligands; (iv) post-processing of the created ligand bases using QSAR models, as a result of which a library of potential ligands for the present biological target may be obtained.

In this study, we present an integrated computational approach to *in silico* drug design involving theoretical procedures, such as virtual screening, high-throughput docking, molecular dynamics simulations (MD), and binding free energy calculations. The efficiency of this approach is demonstrated by the identification of novel potential anti-HIV-1 agents mimicking potent and broad neutralizing antibodies 10e8, VRC01 and 3074 that target three different functionally conserved regions of the HIV-1 Env trimer.

1.1 Computational Methodology for Drug Development

Computer-based approach to in *silico* drug design used in this study included the following consecutive stages: (i) generation of pharmacophore models representing 3D-arrangements of chemical functionalities that make a small molecule active towards its target; (ii) structure-based virtual screening of chemical libraries to discover new ligands on the basis of biological structures; (iii) molecular docking to predict the structure of the intermolecular complex formed between two or more constituent molecules; (iv) MD simulations of this complex followed by binding free energy calculations to evaluate its structural stability.

Brief information on the computational tools supporting the above stages is given below.

1.2 Generation of Pharmacophore Models

Pharmacophore models for virtual screening of antibody-mimetic candidates were generated in agreement with the first step of the pepMMsMIMIC web tool strategy consisting in the identification of amino-acid residues that play a key role in the protein-protein recognition process [10]. This strategy employs as input the 3D structure of a peptide bound to a protein and suggests which chemical structures are able to mimic the protein–protein recognition of this peptide by both pharmacophore and shape similarity techniques. At the same time, all possible combinations of the residues exhibiting critical structural features in 3D space may be used in generation of the templates to screen virtual compound libraries for novel ligands, which present the best similarity to the specific pharmacophore [10]. Based on this strategy, the hotspots of the antibodies of interest for their interactions with the target proteins were derived from the X-ray crystal structures of these immunoglobulins in complexes with the gp120 (VRC01 [11] and 3074 [12]) and gp41 (10e8 [13]) proteins. As a result, 3D structures of the antibody peptides including residues that greatly contribute to the binding were used as the general pharmacophore models for identification of the antibody-mimetic candidates. To identify small-molecule peptidomimetic candidates, short fragments of the general models that target different critical regions of the binding sites of the HIV-1 Env trimer were used as the additional input data for pepMMsMIMIC.

1.3 Shape and Pharmacophore-Based Virtual Screening

The pharmacophore models generated based on the antibody binding hotspots were screened against a library of 17 million conformers obtained from 3.9 million commercially available chemical structures present in the MMsINC database [14]. Screening of this virtual compound library was carried out by four scoring methods that are used in the current version of pepMMsMIMIC [10] to optimize the selection of the peptide mimetics. The tools of pepMMsMIMIC offer five search procedures including different combinations of two scoring approaches, such as ultrafast shape recognition [15] and pharmacophore fingerprints similarity [16]. All these procedures were used for search of the antibody mimetics. The identified compounds were further screened by high-throughput docking to evaluate the efficacy of their binding to the target proteins.

1.4 Molecular Docking

The X-ray crystal structures of the antibody Fabs in complexes with the gp120 (VRC01 [11] and 3074 [12]) and gp41 (10e8 [13]) proteins were used as the rigid receptors for flexible "blind docking" with compounds from the MMsINC database by Autodock Vina [6]. These structures were prepared by adding hydrogen atoms with the Auto-DockTools software [17]. For all compounds, the docked structures with the highest scores were analyzed to identify the molecules that, similarly to the antibodies, exhibit a strong attachment to the antibody-binding sites of the target proteins. As a result, the complexes of top-ranking compounds with these proteins were selected based on the values of scoring functions.

1.5 Molecular Dynamics Simulations

The MD simulations for the docked structures of the top compounds with the target proteins were performed using Amber 11 with the implementation of the Amber ff10 force field [18]. The ANTECHAMBER module was employed to use the Gasteiger atomic partial charges individually for each of the compounds, and the general AMBER force field [19] was used to prepare the force field parameters. Hydrogen atoms were added to the proteins by the xleap program of the AMBER 11 package [18]. The protonation state of histidine residues was checked visually and if necessary either Nδ or Nϵ protonation was chosen to ensure optimal hydrogen bonding. The systems were solvated using TIP3P water [20] as an explicit solvent and simulated in an octahedron box with periodic boundary conditions. The structures were first energy minimized by 500 steps of the steepest descent algorithm followed by 1000 steps of the conjugate gradient method. The atoms of every docked structure were then restrained by an additional harmonic potential with the force constant equal to 1.0 kcal/mol and then heated from 0 to 310 K over 1 ns using a constant volume of the unit cell. Additional equilibration was performed over 1 ns by setting the system pressure to 1.0 atm and by using a weak coupling of the system temperature to a 310 K bath with 2.0 ps characteristic time. Finally, the constraints on the complex assembly were removed and the system was equilibrated again at 310 K over 2 ns under constant volume conditions. After equilibration, the isothermal-isobaric MD simulation (T = 310 K, P = 1.0 atm) generated 30 ns trajectory using a Berendsen barostat with 2.0 ps characteristic time, a Langevin thermostat with collision frequency 2.0 ps^{-1}, a non-bonded cut-off distance of 8 Å, and a simple leapfrog integrator [18] with a 2.0 fs time step and bonds with hydrogen atoms constrained by the SHAKE algorithm [21].

1.6 Binding Free Energy Calculations

The free energy of binding was calculated in AMBER 11 by the MM/PBSA method [22]. Five hundred snapshots were selected from the last 25 ns to estimate the binding free energy, by keeping the snapshots every 50 ps. The polar solvation energies were computed in continuum solvent using Poisson-Boltzmann and ionic strength of 0.1. The non-polar terms were estimated using solvent accessible surface areas.

Based on the MM/PBSA analyses of the MD trajectories, chemical compounds that showed negative free energies of the binding to the target proteins were selected for the final analysis.

2 Results

In the case of bNAb 10e8 that neutralizes up to 98% of diverse HIV-1 strains [13], virtual screening of the MMsINC database combined with molecular docking and MD simulations identified eight top hits that exposed the high-affinity binding to gp41 by targeting the membrane proximal external region (MPER) of this HIV-1 protein, allowing one to consider these molecules as promising peptidomimetic candidates of bNAb 10e8 [23]. Chemical structures of these compounds are shown in Fig. 1.

Figure 2 casts light on the docked structures of the identified compounds (Fig. 1) with the gp41 MPER peptide. In particular, analysis of the MMs03555010-gp41 docked structure indicates (Fig. 2a) that, similarly to bNAb 10e8, this molecule targets the central hinge region of the MPER peptide providing the conformational flexibility necessary for the Env-mediated hemifusion and fusion processes [13, 25].

The docked structures of the identified compounds with gp41 do not undergo substantial rearrangements during the MD simulations, in agreement with the low averages of free energy of their formation that are -16.1 ± 3.4 kcal/mol (MMs03555010), -12.0 ± 4.8 kcal/mol (MMs03769994), -7.3 ± 4.8 kcal/mol (MMs01288397), -7.2 ± 3.9 kcal/mol (MMs02374310), -6.6 ± 4.4 kcal/mol (MMs03064646), -5.7 ± 4.0 kcal/mol (MMs03534576), -5.4 ± 3.6 kcal/mol (MMs01100460), and -5.3 ± 4.5 kcal/mol (MMs00760407).

Analysis of the superimposed complexes of the MPER peptide with the 10e8 Fab and peptidomimetic candidates indicates (Fig. 3) that the identified compounds partially mask the region of gp41 that is targeted by bNAb 10e8. These small molecules bind to the vulnerable spots of this gp41 region and may therefore exhibit the functional mimicry of 10e8. The molecules of interest mimic segment Trp-33, Gly-52c, Pro-52b, Glu-53, Lys-97 of the 10e8 heavy chain (Fig. 3) that forms the direct intermolecular contacts with the functionally important residues of gp41.

The above computer-based technologies have been also applied to the search for potential peptidomimetics of anti-HIV-1 bNAb VRC01 [26] neutralizing over 90% of diverse HIV-1 strains by specific interactions with the CD4-binding site of the Env protein gp120 [11]. As a result, six chemical compounds from MMsINC were shown to exhibit a high affinity to this gp120 site responsible for the HIV-1 attachment to cellular receptor CD4 [26]. The docked models of these compounds with the gp120 core demonstrate intermolecular interactions involving the residues of gp120 important for the HIV-1 binding to CD4. In addition, these complexes show relative stability within the MD simulations that is validated by the values of binding free energy and corresponding standard deviations. These values were shown to be lower than that of

−9.5 ± 0.1 kcal/mol which was determined for the gp120/CD4 complex using isothermal titration calorimetry [26].

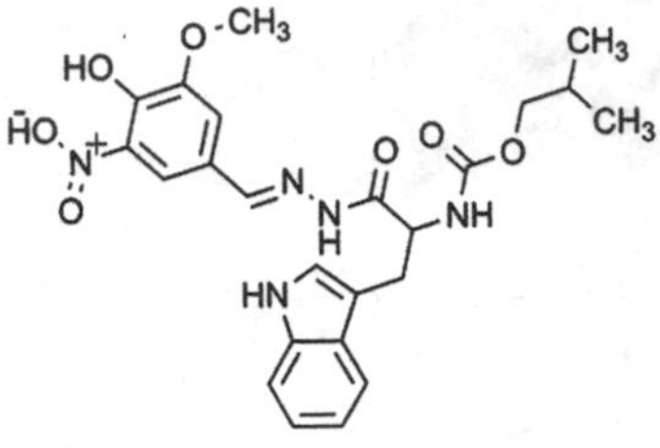

MMs03555010

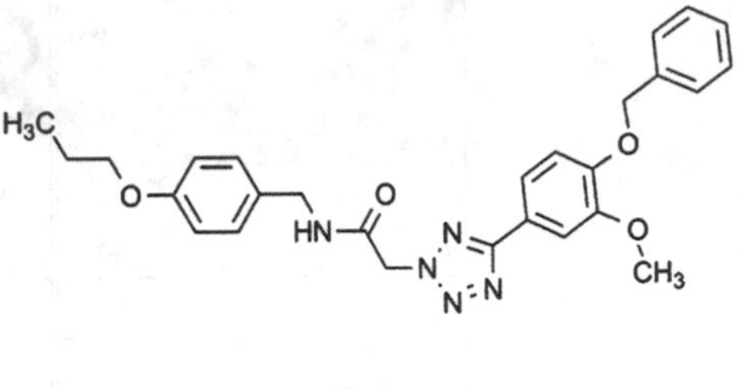

MMs03769994

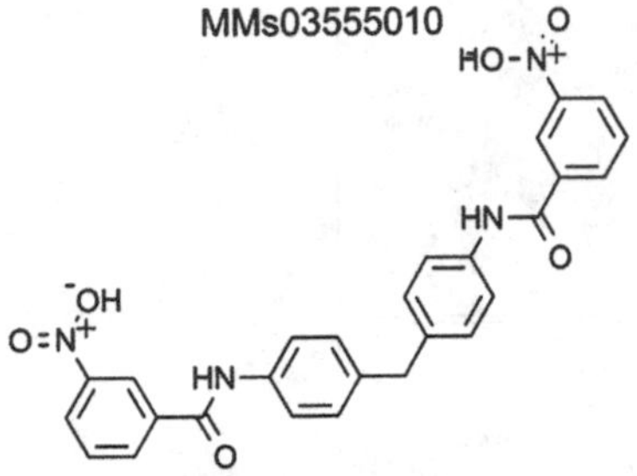

MMs01288397

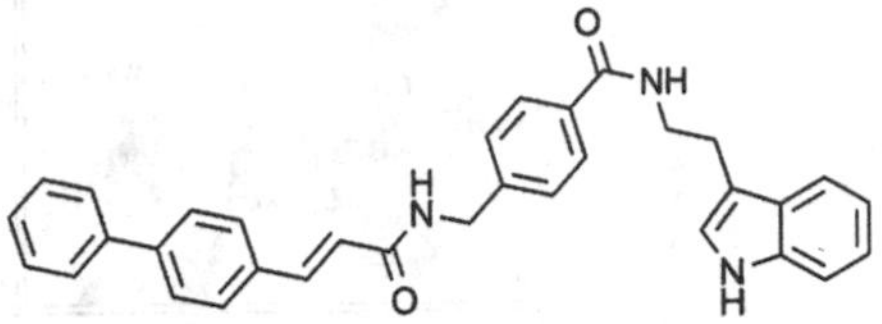

MMs02374310

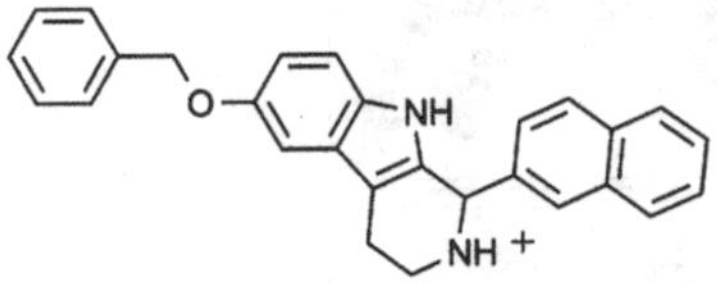

MMs03064646

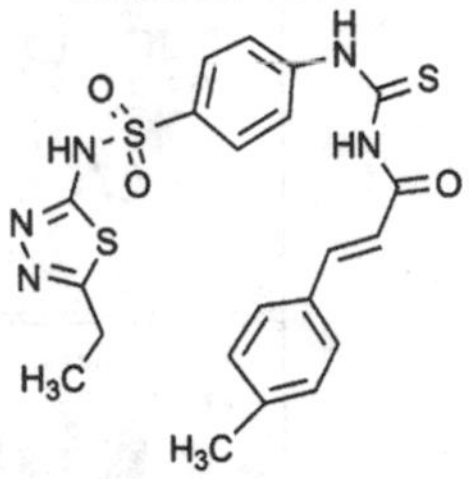

MMs03534576

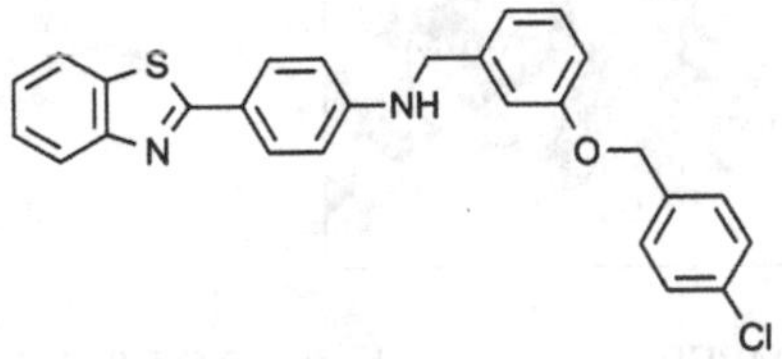

MMs01100460

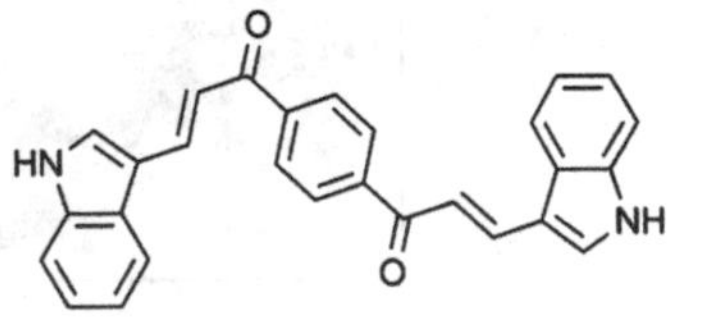

MMs00760407

Fig. 1. Chemical structures of the most probable peptidomimetics of bNAb 10e8. The molecule codes are from the MMsINC database [14]. For these molecules, the four Lipinski's parameters - molecular weight, lipophilicity (LogP), number of H-bond donors and acceptors [24] - are respectively: 497.5 Da, 3.97, 4 and 5 (MMs03555010); 487.6 Da, 3.15, 1 and 7 (MMs03769994); 496.5 Da, 2.36, 2 and 2 (MMs01288397); 499.6 Da, 4.52, 3 and 2 (MMs02374310), 405.5 Da, 0.0, 1 and 1 (MMs03064646); 487.6 Da, 3.62, 3 and 6 (MMs03534576); 457.0 Da, 3.0, 1 and 2 (MMs01100460); 416.5 Da, 2.94, 2 and 2 (MMs00760407)

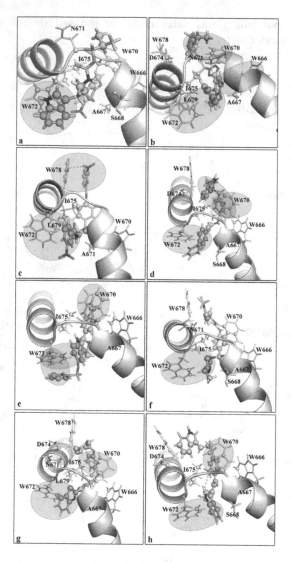

Fig. 2. The docked structures of the gp41 MPER peptide with the MMs03555010, MMs03769994, MMs01288397, MMs02374310, MMs03064646, MMs03534576, MMs01100460, and MMs00760407 compounds. Structures of these compounds are represented by a stick-ball-stick model. The residues of gp41 forming hydrogen bonds, π-π stacking and van der Waals contacts with the 10e8-mimetic candidates are indicated. Structural elements of gp41 and ligands involved in specific π-π interactions are located inside the circles. Hydrogen bonds are shown by dotted lines

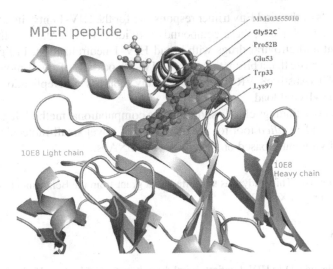

Fig. 3. Superimposed complexes of the HIV-1 MPER peptide with the 10e8 Fab and peptidomimetic candidate MMs03555010. Structure of the 10e8 peptidomimetic is shown by a stick-ball-stick model. Only MMs03555010 in complex with gp41 is shown for peptidomimetic candidates; the complexes of gp41 with the other identified compounds coincide with this supramolecular structure. A Corey-Pauling-Koltun model is used to highlight the 10e8 residues that are mimicked by the identified compounds

In contrast to VRC01 targeting CD4-binding site of gp120 [11], bNAb 3074 binds to the third variable (V3) loop of gp120 [12]. This domain of gp120 plays a central role in the biology of the HIV-1 Env trimer as a principal target for neutralizing antibodies and as a major determinant in the switch from the non-syncytium-inducing to the syncytium-inducing form of HIV-1 that is associated with accelerated disease progression [9]. Application of the above methodology to the search for the most probable mimetic candidates of 3074 resulted in the identification of four top hits demonstrating a strong attachment to the HIV-1 V3 loop [27]. Specific binding to the V3 loop was shown to be accomplished primarily by π-π interactions between the aromatic rings of the peptidomimetics and the conserved Phe-20 and/or Tyr-21 of the V3 immunogenic crown [27]. In a mechanism similar to that of bNAb 3074, these compounds were found to block the tip of the V3 loop forming its invariant structural motif that contains residues critical for the HIV-1 binding to a chemokine co-receptor, either CCR5 or CXCR4, which is required for viral entry. With these findings, the following conclusion was made: the compounds selected form promising scaffolds for the rational design of novel, potent, and broad-spectrum anti-HIV-1 therapeutics [27].

3 Conclusions

An integrated computational approach used in this study allowed one to identify eighteen top hits able to mimic anti-HIV-1 bNAbs 10e8 (eight compounds), VRC01 (six compounds), and 3074 (four compound). These small molecules target three different vulnerable spots of the viral Env trimer responsible for the HIV-1 entry into target cells. Based on the data obtained, these compounds provide good scaffolds for the development of potent antiretroviral drugs with broad HIV-1 neutralization. In light of these data, we also suppose that a bifunctional anti-HIV-1 "cocktail" of small-molecule peptidomimetics of bNAbs 10e8, VRC01, and 3074 may suppress viral replication and reduce the plasma HIV-1 viral load.

Thus, the above findings clearly show that the computational methodology presented here is a powerful *in silico* tool for the discovery of novel small-molecule functional antagonists of viral entry based on the anti-HIV-1 bNAbs.

Acknowledgements. This study was supported by a grant from the Belarusian Foundation for Basic Research (project X15-022).

References

1. Arts, E.J., Hazuda, D.J.: HIV-1 antiretroviral drug therapy. Cold Spring Harb. Perspect. Med. **2**, a007161 (2012)
2. Kumari, G., Singh, R.K.: Highly active antiretroviral therapy for treatment of HIV/AIDS patients: current status and future prospects and the Indian scenario. HIV AIDS Rev. **11**, 5–14 (2012)
3. Wilen, C.B., Tilton, J.C., Doms, R.W.: HIV: cell binding and entry. Cold Spring Harb. Perspect. Med. **2**, a006866 (2012)
4. Acharya, P., et al.: HIV-1 gp120 as a therapeutic target: navigating a moving labyrinth. Expert Opin. Ther. Targets **19**, 1–19 (2015)
5. Courter, J.R., et al.: Structure-based design, synthesis and validation of CD4-mimetic small molecule inhibitors of HIV-1 entry: conversion of a viral entry agonist to an antagonist. Acc. Chem. Res. **47**, 1228–1237 (2014)
6. McCoy, L.E., Weiss, R.A.: Neutralizing antibodies to HIV-1 induced by immunization. J. Exp. Med. **210**, 209–223 (2013)
7. Mascola, J.R., Haynes, B.F.: HIV-1 neutralizing antibodies: understanding nature's pathways. Immunol. Rev. **254**, 225–244 (2013)
8. Haynes, B.F., McElrath, M.J.: Progress in HIV-1 vaccine development. Curr. Opin. HIV AIDS **8**, 326–332 (2013)
9. Andrianov, A.M.: Human immunodeficiency virus-1 gp120 V3 loop for anti-acquired immune deficiency syndrome drug discovery: computer-aided approaches to the problem solving. Expert Opin. Drug Discov. **6**, 419–435 (2011)
10. Floris, M.J., et al.: Swimming into peptidomimetic chemical space using pepMMsMIMIC. Nucleic Acids Res. **39**, W261–W269 (2011)
11. Zhou, T., et al.: Structural basis for broad and potent neutralization of HIV-1 by antibody VRC01. Science **329**, 811–817 (2010)
12. Jiang, X., et al.: Conserved structural elements in the V3 crown of HIV-1 gp120. Nature Struct. Mol. Biol. **17**, 955–961 (2010)
13. Huang, J., et al.: Broad and potent neutralization of HIV-1 by a gp41-specific human antibody. Nature **491**, 406–412 (2012)
14. Masciocchi, J., et al.: MMsINC: a large-scale chemoinformatics database. Nucleic Acids Res. **37**, D284–D290 (2009)

15. Ballester, P.J., Richards, W.G.: Ultrafast shape recognition to search compound databases for similar molecular shapes. J. Comput. Chem. **28**, 1711–1723 (2007)
16. Mason, J.S., et al.: New 4-point pharmacophore method for molecular similarity and diversity applications: overview of the method and applications, including a novel approach to the design of combinatorial libraries containing privileged substructures. J. Med. Chem. **42**, 3251–3264 (1999)
17. Trott, O., Olson, A.J.: Software news and update AutoDock Vina: Improving the speed and accuracy of docking with a new scoring function, efficient optimization, and multithreading. J. Comput. Chem. **31**, 455–461 (2010)
18. Case, D.A., et al.: AMBER 11. University of California, San Francisco (2010)
19. Wang, R.M., et al.: Development and testing of a general Amber force field. J. Comput. Chem. **25**, 1157–1174 (2004)
20. Jorgensen, W.L., et al.: Comparison of simple potential functions for simulating liquid water. J. Chem. Phys. **79**, 926–935 (1983)
21. Ryckaert, J.P., Ciccotti, G., Berendsen, H.J.C.: Numerical integration of the Cartesian equations of motion of a system with constraints: molecular dynamics of n-alkanes. J. Comput. Phys. **23**, 327–341 (1977)
22. Massova, I., Kollman, P.A.: Computational alanine scanning to probe protein-protein interactions: a novel approach to evaluate binding free energies. J. Am. Chem. Soc. **121**, 8133–8143 (1999)
23. Andrianov, A.M., Kashyn, I.A., Tuzikov, A.V.: Identification of novel HIV-1 fusion inhibitor scaffolds by virtual screening, high-throughput docking and molecular dynamics. JSM Chem. **4**(2), 1022 (2016)
24. Lipinski, C.A., et al.: Lead- and drug-like compounds: the rule-of-five revolution. Adv. Drug Deliv. Rev. **46**, 3–26 (2001)
25. Cheng, Y.: Elicitation of antibody responses against the HIV-1 gp41 Membrane Proximal External Region (MPER). Doctoral dissertation, Harvard University (2014). (http://nrs.harvard.edu/urn-3:HUL.InstRepos:12269838)
26. Andrianov, A.M., Kashyn, I.A., Tuzikov, A.V.: Computational discovery of novel HIV-1entry inhibitors based on potent and broad neutralizing antibody VRC01. J. Mol. Graph. Model. **61**, 262–271 (2015)
27. Andrianov, A.M., Kashyn, I.A., Tuzikov, A.V.: Discovery of novel Anti-HIV-1 agents based on a broadly neutralizing antibody against the envelope gp120 V3 loop: a computational study. J. Biomol. Struct. Dyn. **32**, 1993–2004 (2014)

Pattern Recognition and Image Analysis

Dynamic Colour Clustering for Skin Detection Under Different Lighting Conditions

Nadia Brancati[✉], Giuseppe De Pietro, Maria Frucci, and Luigi Gallo

Institute for High-Performance Computing and Networking,
National Research Council (ICAR-CNR), 80131 Naples, Italy
{nadia.brancati,giuseppe.depietro,maria.frucci,
luigi.gallo}@cnr.it

Abstract. Skin detection is an important process in many applications like hand gesture recognition, face detection and ego-vision systems. This paper presents a new skin detection method based on a dynamic generation of the skin cluster range in the *YCbCr* color space, by taking into account the lighting conditions. The method is based on the identification of skin color clusters in the *YCb* and *YCr* subspaces. The experimental results, carried out on two publicly available databases, show that the proposed method is robust against illumination changes and achieves satisfactory results in terms of both qualitative and quantitative performance evaluation parameters.

Keywords: Skin detection · Dynamic clustering · YCbCr colour space

1 Introduction

Skin detection is an important issue in color image processing, which has been extensively studied over the years. It is a useful technique for the detection, segmentation and tracking of human skin in images or video streams. The interest in skin detection algorithms derives from their applicability to a wide range of applications such as gesture recognition, video surveillance, human computer interaction, ego-vision systems, human activity recognition [4, 6, 21], hand gestures detection and tracking [7, 8, 23, 31], nude images and video blocking [5, 26], feature extraction for content-based image retrieval [20], and age estimation [22].

Skin detection is a process that allows the extraction of candidate skin pixels in an image. In most cases, skin detection is performed by using pixel based techniques: a pixel is classified as a skin or non-skin pixel, independently from its neighbors, and only by using pixel color information. In addition, region based skin segmentation methods make use of extra information, for example spatial arrangement or texture information on the pixels detected in the skin detection process, to determine the boundaries of human skin regions [16, 17]. Therefore, a good pixel based method for skin detection can narrow the computational cost of the next process of segmentation, and, moreover, can improve the results of the segmentation. The main issue is to achieve a satisfactory skin detection under uncontrolled lighting conditions, since many applications, for example egovision systems, require the detection of skin human regions, both indoors and outdoors, with

© Springer International Publishing AG 2017
V.V. Krasnoproshin and S.V. Ablameyko (Eds.): PRIP 2016, CCIS 673, pp. 27–35, 2017.
DOI: 10.1007/978-3-319-54220-1_3

high or low illumination conditions. Most approaches use specific color spaces to decorrelate chromatic components from luminance, since they are less sensitive to lighting conditions [9, 12]. However, some studies [10, 14] have shown that the luminance component plays an important role in skin detection and so it should not be rejected.

In the present work, an explicit skin cluster method, in the $YCbCr$ colour space, is proposed. The method takes into account the illumination changes of the examined image, and tries to minimize both false positives and false negatives. It results to be computationally efficient for real-time applications.

The rest of the paper is organized as follows: in Sect. 2, a description of the related work for skin detection is presented; Sect. 3 describes the proposed approach; in Sect. 4, some results and comparative evaluations performed on two publicly available databases are reported; finally, in Sect. 5 conclusions are drawn.

2 Related Work

Many recent surveys describe the various skin detection approaches [13, 18, 30]. Many approaches have been proposed for the skin colour detection; they include linear classifiers [9, 11, 12, 19, 25, 27], Bayesian [10, 15] or Gaussian classifiers [11, 29], and artificial neural network [1, 2, 28].

It has been demonstrated that the human skin colour can be modelled in many colour spaces [3, 24].

In [19], heuristic rules in the RGB color space are used to detect skin pixels; these rules depending on the image illumination; other methods adopt linear or non-linear transformations of the RGB colour space, such that other colour spaces can be generated. In particular, those colour spaces that separate the luminance and the chrominance components, are the most commonly used in skin colour detection approaches. This is the case of HSV and $YCbCr$, which are a non- linear and a linear transformation of the RGB colour space, respectively. Concerning to the HSV colour space, Hue (H) and Saturation (S) are the chrominance components, while Value (V) is the luminance component. Most of the methods that work in this colour space ignore the luminance component, since it does not result to be discriminant [12, 25]; however, also some methods that include the luminance in the process of skin detection have been presented [27]. Concerning to the $YCbCr$ colour space the chrominance components Cb and Cr are obtained by subtracting the luminance component Y from blue and from red, respectively. Also in this case, some approaches ignore the luminance component [9], while others take it into account [14]. However, it has been demonstrated that skin colour is non-linearly dependent on the luminance component in different colour spaces, thus the luminance component should be included in the skin detection process [11, 14].

Explicit cluster methods are based on the definition of colour rules, in particular on the definition of a colour range for skin pixels [9, 25], or on the definition of a shape for skin pixel distribution (e.g., rectangle and ellipse). In [14], two different skin cluster models, that take into account the luminance component, have been proposed. In the first model, the skin clusters are determined by two central curves, one for the YCr and one for the YCb subspace, and by their spreads in the respective subspaces. In the second model, a single skin cluster is represented by an ellipse in a transformed $CbCr$ subspace.

In this paper, a new approach, that works in the $YCbCr$ colour space is proposed. In particular, taking into account the illumination conditions of the examined image, a dynamic cluster for the YCb and YCr subspaces is computed.

3   The Proposed Approach

As already shown in [14], the distribution of skin pixels in the YCb and YCr subspaces presents a trapezoidal shape (see Fig. 1a), differently from distribution of skin and non-skin pixels (see Fig. 1b).

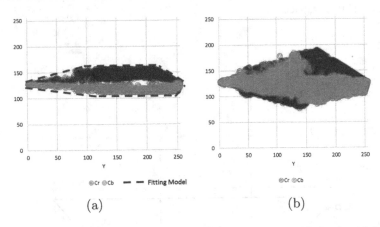

(a)   (b)

Fig. 1. Cr and Cb components as function of Y component for a specific image: (a) distribution of skin pixels; (b) distribution of skin and non-skin pixel.

Moreover, we experimentally observed that the size and shape of these trapezia change depending on the lighting conditions. In particular, we have observed that:

– for images in high illumination conditions, the bases of the two trapezia in the YCb and YCr subspaces representing the skin colour clusters are larger than those associated with the skin colour clusters in low illumination conditions;
– the positions of the vertices of the trapezia change according to the illumination conditions of the examined image;
– for skin pixels, the minimum value of Cr (in the following Cr_{min}) and the maximum value of Cb (in the following Cb_{max}) are practically fixed at the values 133 and 128, respectively, as reported in [9], while the maximum value of Cr and the minimum value of Cb strongly change with the illumination conditions;
– for a skin pixel, the values of the Cr and Cb generally satisfy the following conditions:

$$133 \leq Cr \leq 183$$
$$77 \leq Cb \leq 128$$

With reference to the Fig. 2, the vertices A and D of the larger basis of the trapezium related to the YCr skin subspace are given by (Y_{min}, Cr_{min}) and (Y_{max}, Cr_{min}), where $Y_{min} = 0$, $Y_{max} = 255$ and $Cr_{min} = 133$. The same applies to the vertices E and H of the larger basis of the trapezium related to the YCb skin subspace that are given by (Y_{min}, Cb_{max}) and (Y_{max}, Cb_{max}), with $Cb_{max} = 128$. Concerning to the vertices B and C of the shorter basis of the trapezium associated with the YCr skin subspace, they are set to (Y_0, Cr_{max}) and (Y_1, Cr_{max}). Taking into account the histogram of the pixels with values of Cr in the range $[133,183]$, Cr_{max} is set to the maximum of Cr, associated with at least a 10% of image pixels. So, Y_0 and Y_1 values are set as the 5th percentile and the 95th percentile of the Y component, respectively, considering all the pixels of the image with $Cr = Cr_{max}$. The same process is applied to find the vertices F and G with coordinates (Y_2, Cb_{min}) and (Y_3, Cb_{min}) respectively, of the shorter basis of the trapezium associated with the YCb skin subspace.

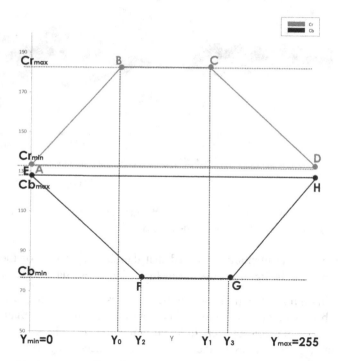

Fig. 2. Graphical representation of $Y_{min}, Y_{max}, Y_0, Y_1, Y_2, Y_3, Cr_{max}, Cr_{min}, Cbmax, Cbmin$.

Set a Y value, a point on the upper border of the trapezium in the YCr subspace will have coordinates $(Y, T_{Cr}(Y))$, while a point on the lower bound of the trapezium in the YCb subspace, will have coordinates $(Y, T_{Cb}(Y))$. $T_{Cr}(Y)$ and $T_{Cb}(Y)$ are given by:

$$T_{Cr}(Y) = \begin{cases} Cr_{min} + d_{Cr}\dfrac{Y - Y_{min}}{Y_o - Y_{min}} & Y \in [Y_{min}, Y_0] \\ Cr_{max} & Y \in [Y_0, Y_1] \\ Cr_{max} - d_{Cr}\dfrac{Y - Y_1}{Y_{max} - Y_1} & Y \in [Y_1, Y_{max}] \end{cases}$$

where $d_{Cr} = Cr_{max} - Cr_{min}$

$$T_{Cb}(Y) = \begin{cases} Cb_{max} - d_{Cb}\dfrac{Y - Y_{min}}{Y_2 - Y_{min}} & Y \in [Y_{min}, Y_2] \\ Cb_{min} & Y \in [Y_2, Y_3] \\ Cb_{min} + d_{Cb}\dfrac{Y - Y_3}{Y_{max} - Y_3} & Y \in [Y_3, Y_{max}] \end{cases}$$

where $d_{Cb} = Cb_{max} - Cb_{min}$

Finally, we classify a pixel as skin pixel, if it satisfies the following two conditions:

$$Cr(Y) \in \left[Cr_{min}, T_{Cr}(Y)\right]$$
$$AND$$
$$Cb(Y) \in \left[T_{Cb}(Y), Cb_{max}\right]$$

4   Results and Comparison

The proposed approach has been compared with the method described in [9], which also works in the $YCbCr$ colour space but with a fixed colour range, and with the method presented in [14], considering the both formulations of the skin cluster models. The approach has been tested on the Hand Gesture Recognition (HGR) database [17], containing 1,558 skin images of human hand and arm postures taken with different lighting conditions and on the Compaq database [15], a large database that consists of 4,675 colour images, containing skin images in unconstrained illumination and background conditions. Some qualitative results, for our approach and for the methods with which we compared, are shown in Figs. 3 and 4, for some selected images of the HGR and Compaq databases. Starting from a qualitative analysis of the results, it is clear that the method proposed in [9] generally obtains good results, but in some cases, many false positives are found (see row 3 in Fig. 3 and rows 1 and 3 in Fig. 4); in fact, the pixels belonging to regions of eyes and mouth or of background are generally wrongly detected as skin pixels. Concerning to the method in [14], in the $YCbCr$ formulation, also in this case, many false positives are found, particularly in presence of high or low illumination conditions (see all the results of the Fig. 3 and rows 2 and 5 in Fig. 4); the skin detection performance improves on the case of its formulation in the transformed $CbCr$ subspace, but in some cases many false negative are detected (see row 1 in Fig. 3 and rows 1 and 4 in Fig. 4).

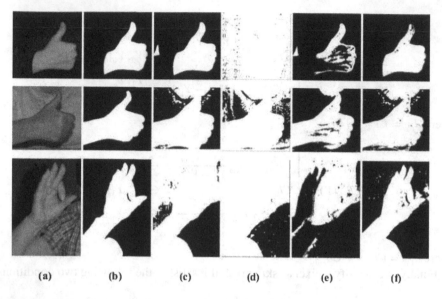

Fig. 3. Qualitative analysis of skin detection results on the HGR database: (a) the input image; (b) the ground truth; (c) Chai, Ngan [9]; (d) Hsu et al. in the *YCbCr* space [14]; (e) Hsu et al. in the *CbCr* subspace [14]; and (f) the proposed method.

Moreover, quantitative results in terms of F-measure are reported in Table 1, for all the analysed approaches. The proposed approach outperforms the other methods in terms of F-measure.

Table 1. F-measure for the skin detection approaches on the Compaq and HGR databases.

Method	Compaq	HGR
CbCr [9]	0.4373	0.7967
YCbCr [14]	0.3104	0.3078
CbCr [14]	0.4312	0.5814
Proposed method	**0.4649**	**0.8295**

Finally, the computational cost of the proposed approach has been estimated. The performance of the algorithm has been estimated on a PC equipped with an Intel Xeon E5-2623 at 3 GHz, and with 16 GB RAM. For an image with a size of 320×480, the execution time is, on average, 8 ms.

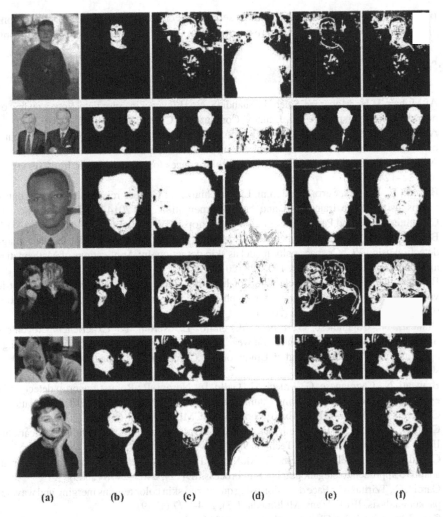

Fig. 4. Qualitative analysis of skin detection results on the Compaq database: (a) the input image; (b) the ground truth; (c) Chai, Ngan [9]; (d) Hsu et al. in the *YCbCr* space [14]; (e) Hsu et al. in the *CbCr* subspace [14]; and (f) the proposed method.

5   Conclusion

We have presented a new approach for skin detection in the *YCbCr* color space. The method shows some robustness to variations in illumination conditions, because the skin cluster range in the *YCbCr* color space is defined dynamically, taking into account the luminance component. In particular, two clusters are found, one in the *YCb* subspace and one in the *YCr* subspace.

The performance of the method has been tested on two publicly available databases, producing satisfactory results both qualitatively and in terms of quantitative performance evaluation parameters such as F-measure. The results of a comparative analysis are promising. With respect to methods based on fixed cluster ranges, the proposed one provide adequate results also on images acquired in low or high illumination conditions.

References

1. Al-Mohair, H.K., Mohamad-Saleh, J., Suandi, S.A.: Human skin color detection: a review on neural network perspective. Int. J. Innov. Comput. Inf. Control **8**(12), 8115–8131 (2012)
2. Al-Mohair, H.K., Saleh, J.M., Suandi, S.A.: Hybrid human skin detection using neural network and k-means clustering technique. Appl. Soft Comput. **33**, 337–347 (2015)
3. Albiol, A., Torres, L., Delp, E.J.: Optimum color spaces for skin detection. In: ICIP, vol. 1, pp. 122–124 (2001)
4. Baraldi, L., Paci, F., Serra, G., Benini, L., Cucchiara, R.: Gesture recognition in ego-centric videos using dense trajectories and hand segmentation. In: 2014 IEEE Conference on Computer Vision and Pattern Recognition Workshops (CVPRW), pp. 702–707. IEEE (2014)
5. Basilio, J.A.M., Torres, G.A., Pérez, G.S., Medina, L.K.T., Meana, H.M.P.: Explicit image detection using YCbCr space color model as skin detection. In: Applications of Mathematics and Computer Engineering, pp. 123–128 (2011)
6. Betancourt, A.: A sequential classifier for hand detection in the framework of egocentric vision. In: 2014 IEEE Conference on Computer Vision and Pattern Recognition Workshops (CVPRW), pp. 600–605. IEEE (2014)
7. Brancati, N., Caggianese, G., Frucci, M., Gallo, L., Neroni, P.: Experiencing touchless interaction with augmented content on wearable head mounted displays in cultural heritage applications. Personal and Ubiquitous Computing https://dx.doi.org/10.1007/s00779-016-0987-8
8. Brancati, N., Caggianese, G., Frucci, M., Gallo, L., Neroni, P.: Robust fingertip detection in egocentric vision under varying illumination conditions. In: 2015 IEEE International Conference on Multimedia & Expo Workshops (ICMEW), pp. 1–6. IEEE (2015)
9. Chai, D., Ngan, K.N.: Face segmentation using skin-color map in videophone applications. IEEE Trans. Circ. Syst. Video Technol. **9**(4), 551–564 (1999)
10. Cheddad, A., Condell, J., Curran, K., McKevitt, P.: A skin tone detection algorithm for an adaptive approach to steganography. Sig. Process. **89**(12), 2465–2478 (2009)
11. Garcia, C., Tziritas, G.: Face detection using quantized skin color regions merging and wavelet packet analysis. IEEE Trans. Multimedia **1**(3), 264–277 (1999)
12. Guo, J.M., Liu, Y.F., Chang, C.H., Nguyen, H.S.: Improved hand tracking system. IEEE Trans. Circ. Syst. Video Technol. **22**(5), 693–701 (2012)
13. Hjelmås, E., Low, B.K.: Face detection: A survey. Comput. Vis. Image Underst. **83**(3), 236–274 (2001)
14. Hsu, R.L., Abdel-Mottaleb, M., Jain, A.K.: Face detection in color images. IEEE Trans. Pattern Anal. Mach. Intell. **24**(5), 696–706 (2002)
15. Jones, M.J., Rehg, J.M.: Statistical color models with application to skin detection. Int. J. Comput. Vision **46**(1), 81–96 (2002)
16. Kawulok, M.: Fast propagation-based skin regions segmentation in color images. In: 2013 10th IEEE International Conference and Workshops on Automatic Face and Gesture Recognition (FG), pp. 1–7. IEEE (2013)
17. Kawulok, M., Kawulok, J., Nalepa, J.: Spatial-based skin detection using discriminative skin-presence features. Pattern Recogn. Lett. **41**, 3–13 (2014)
18. Khan, R., Hanbury, A., Stöttinger, J., Bais, A.: Color based skin classification. Pattern Recogn. Lett. **33**(2), 157–163 (2012)

19. Kovac, J., Peer, P., Solina, F.: Human Skin Color Clustering For Face Detection, vol. 2. IEEE (2003)
20. Kruppa, H., Bauer, Martin, A., Schiele, B.: Skin patch detection in real-world images. In: Gool, L. (ed.) DAGM 2002. LNCS, vol. 2449, pp. 109–116. Springer, Heidelberg (2002). doi: 10.1007/3-540-45783-6_14
21. Li, C., Kitani, K.M.: Pixel-level hand detection in ego-centric videos. In: 2013 IEEE Conference on Computer Vision and Pattern Recognition (CVPR), pp. 3570–3577. IEEE (2013)
22. Luu, K., Bui, T.D., Suen, C.Y., Ricanek, K.: Spectral regression based age determination. In: 2010 IEEE Computer Society Conference on Computer Vision and Pattern Recognition Workshops (CVPRW), pp. 103–107. IEEE (2010)
23. Nalepa, J., Grzejszczak, T., Kawulok, M.: Wrist localization in color images for hand gesture recognition. Man-Machine Interactions 3, 79–86 (2014). Springer
24. Shin, M.C., Chang, K.I., Tsap, L.V.: Does colorspace transformation make anydifference on skin detection? In: Proceedings of the Sixth IEEE Workshop on Applications of Computer Vision 2002 (WACV 2002), pp. 275–279. IEEE (2002)
25. Sobottka, K., Pitas, I.: A novel method for automatic face segmentation, facial feature extraction and tracking. Sig. Process. Image Commun. 12(3), 263–281 (1998)
26. Stöttinger, J., Hanbury, A., Liensberger, C., Khan, R.: Skin paths for contextual flagging adult videos. In: Bebis, G., Boyle, R., Parvin, B., Koracin, D., Kuno, Y., Wang, J., Pajarola, R., Lindstrom, P., Hinkenjann, A., Encarnação, Miguel, L., Silva, Cláudio, T., Coming, D. (eds.) ISVC 2009. LNCS, vol. 5876, pp. 303–314. Springer, Heidelberg (2009). doi: 10.1007/978-3-642-10520-3_28
27. Tsekeridou, S., Pitas, I.: Facial feature extraction in frontal views using biometric analogies. In: 9th European Signal Processing Conference (EUSIPCO 1998), pp. 1–4. IEEE (1998)
28. Yogarajah, P., Condell, J., Curran, K., McKevitt, P., Cheddad, A.: A dynamic threshold approach for skin tone detection in colour images. Int. J. Biometrics 4(1), 38–55 (2011)
29. Zaidan, A., Ahmad, N.N., Karim, H.A., Larbani, M., Zaidan, B., Sali, A.: Image skin segmentation based on multi-agent learning bayesian and neural network. Eng. Appl. Artif. Intell. 32, 136–150 (2014)
30. Zhao, W., Chellappa, R., Phillips, P.J., Rosenfeld, A.: Face recognition: A literature survey. ACM Comput. Surv. (CSUR) 35(4), 399–458 (2003)
31. Zhu, Y., Xu, G., Kriegman, D.J.: A real-time approach to the spotting, representation, and recognition of hand gestures for human–computer interaction. Comput. Vis. Image Underst. 85(3), 189–208 (2002)

New Fast Content Based Skew Detection Algorithm for Document Images

Mohd Amir and Abhishek Jindal[✉]

Newgen Software Technologies Ltd., A-6 Satsang Vihar Marg, Qutab Institutional Area,
New Delhi, 110067, India
{mohd.amir,abhishek.jindal}@newgen.co.in
http://newgensoft.com/

Abstract. The process of obtaining digitized copies of documents has become an indispensable part of many businesses around the world, including but not limited to Banking, Insurance and Telecom domain. The document capturing part is usually done with the help of scanning devices that have evolved from being fixed flat-bed ones' to hand-held devices. While capturing documents using the ADF scanners, skew related defects usually get inserted in the images. These defects degrade the legibility of the documents and hence the accuracy of the OCR and ICR extraction engines. The degraded accuracy of the extraction engines inversely affects the businesses that capture critical user information. Thus, it becomes necessary to correct such errors. Since the skew detection part holds a small but significant part in the whole process of form processing, it has to work accurately and efficiently to enhance performance of complete system. Here, a new faster content based approach for skew detection on document images is proposed. The proposed approach considers only the printed text present in the image, for skew detection, thus eliminating the errors arising due to the pictorial information, noise, scanning artifacts, etc.

Keywords: Component labelling · Document deskewing · Skew angle · OCR · Skew detection · Document analysis · Line fitting · Deskew

1 Introduction

Globalization and increasing world population together have led to an increase in the customer-base of companies worldwide. Paper-based application forms, being a popular medium for customer data capturing, have significantly increased the amount of documents that need to be digitized for processing. Hence, the process of obtaining digitized copies of documents has become an indispensable part of many businesses around the world. The document capturing part is usually done with the help of scanning devices that range from fixed flatbed scanners to hand held scanners and smart-phones. Moreover, the scanning may be either manual using flatbed scanners or batch scanning through ADF scanners, depending on the amount of digitization requirement. Batch scanning requires collection of documents into batches that are fed to the feeder scanners for scanning. The process of batch creation and sometimes even the scanner hardware leads to insertion of skew related defects in the scanned document images. But in case of huge digitization requirements this process cannot

© Springer International Publishing AG 2017
V.V. Krasnoproshin and S.V. Ablameyko (Eds.): PRIP 2016, CCIS 673, pp. 36–43, 2017.
DOI: 10.1007/978-3-319-54220-1_4

be avoided. Manual scanning, on the other hand, is also prone to skew insertion due to various human factors. Hence, scanning or digitization of any type is prone to skew defect insertion. The digitized documents are then fed to the electronic data processing systems for various operations like data extraction, OCR or ICR detection, etc. The efficiency and accuracy of all these operations is dependent on the legibility of the data present in the document images.

The skew inserted in the images reduces (depending on the degree of skew) the legibility of the data. This inversely affects the accuracy of the OCR and extraction engines. The only way to improve OCR engine accuracy on such images is to either rescan the document or correct the document image using some image processing functionalities. The rescanning of document becomes an overhead and is not certain to remove the defect in the rescanned document. There are chances that the same defect might be present in the new image as well. Also, this type of method doesn't work when there are hundreds or thousands of documents to be rescanned. This is where the image processing part plays its role. So, image processing algorithms are applied on the skewed documents to correct them prior to further processing without the need for rescanning. This process of correcting the skew defects of image is called deskewing. It comprises of two phases: firstly identifying the skew angle of the image and then rotating the image with the angle to correct the defect. These functions are applied to the images at the post-production stage using graphics software. Since the skew correction part holds a small but significant part in the whole process of form processing, it has to work accurately and also efficiently to enhance performance of the entire system.

Here, a new faster approach for skew detection on document images, on the basis of the printed text, is being discussed. Most of the algorithms [1,2] available work on the horizontal and vertical projections of image data, sometimes causing the noise and non-textual components deteriorate the skew detection accuracy. The document images or application forms make for a major part of the images being input to the form processing systems. Supporting documents or other pictorial images form a small part of it. Hence, as per our experimental runs on a set of 20400 skewed images, it is best to observe just the printed text present and avoid the noisy and pictorial information. The proposed approach provides 130% better accuracy and 330% better performance over the projection based and line based skew detection approaches.

2   The Proposed Approach

This system for detection of skew from document image can be decomposed into six stages. It involves capturing the document, pre-processing the captured document, detecting the connected components in the preprocessed document, filtering the detected components for processable textual components, identifying skew of the selected component and storing it in a list and finally calculation of the final skew angle.

The proposed approach, even after bearing the overhead of preprocessing functionalities like binarization and resizing, gives better performance in comparison to the projection based approaches. It accurately calculates the skew angle, of document images, in the range of $\pm30°$ (Fig. 1).

Fig. 1. Input image having skew

Variables being used in the paper:

£(t):   the list having calculated skew angles of textual lines
£$_{sorted}$(t): the list having sorted data of £(t)
λ: the length of £$_{sorted}$(t)
η$_1$:   lower skew range in the sorted skew list
η$_2$:   upper skew range in the sorted skew list

A. Capturing the document image

This step involves capturing image of the input document. The algorithm processes mainly the printed text information present in the input image. The image can be captured using any capturing device, but the orientation must be landscape. The reason for operating only on landscape images is that the proposed approach does not determine the orientation of the input image and calculates just the skew error present in it. If an image with portrait orientation is given as input to the algorithm, then many components might get filtered out in the latter stages, thus decreasing the skew detection accuracy of the approach.

B. Preprocessing captured document

In this phase, some preprocessing is done on the captured document image for preparing it for further processing. Preprocessing is done to increase the time performance of the approach by reducing the number of processable components.
    The steps included in the preprocessing phase are:

1. Binarizing input image

    The input image provided as input to the approach can be of any color-depth viz. True color, black and white, 4-bits per pixel image or 8-bits per pixel image. The preprocessing steps require a 1 bit per pixel image as input. Hence the first step of preprocessing includes binarizing the input image. The algorithm utilized for binarization is Otsu's threshold selection method [3]. This method reduces a Gray level image to a binary image. It assumes that the image to be thresholded contains two classes of pixels or bi-modal histogram (e.g. foreground and background), then calculates the optimum threshold separating those two classes so that their combined spread (intra-class variance) is minimal.

2. Resizing binarized image

In this step, the binarized image is resized to 50 DPI to speed up the algorithm and removing unwanted noise. The algorithm used for resizing follows uniform sampling method.

$f(x,y)$ represents the original continuous image, $f_s(m,n)$ the sampled image.

$$f_s(m, n) = f(m\Delta x, n\Delta y), \tag{1}$$

where, $0 \leq m \leq M - 1; 0 \leq n \leq N - 1$.

Fig. 2. Resized image

Δx and Δy are vertical and horizontal sampling intervals. $f_{s,x} = 1/\Delta x$, $f_{s,y} = 1/\Delta y$ are vertical and horizontal sampling frequencies respectively. M and N are the output image dimensions (Fig. 2).

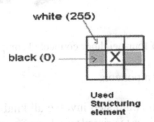

Fig. 3. Dilation matrix

3. Dilating resized image

In this step, the resized image is dilated horizontally by one pixel as shown in Fig. 3 below. Dilation is done to connect the neighboring pixels of textual data together for better detection of printed text properties.

Fig. 4. Dilated image

C. Detection of connected components

In this phase, the preprocessed image is traversed for detection of four-connected components using four-neighbor connected component-labeling algorithm [4]. Figure 5 depicts the sample output of the component labelling algorithm on a small image.

Fig. 5. Output of four-neighbor connected component-labelling

D. Filtering the non-textual components

The components detected by the third step involve all kind of components ranging from small salt-pepper noise and graphics to printed text. For accurate detection of skew angle and faster processing, it is necessary to filter the non-textual components. This step performs this filtering functionality and gives only processable and preprocessed printed textual components for next steps [5, 6] (Fig. 4).

E. Identifying skew of textual components

The skew angle of each component received from the fourth step is calculated after applying line fitting on it.

Steps followed for line fitting of each component:

Step 1: Calculate the sum of the x-values and y-values respectively
Step 2: Compute the sum of the squares of the x-values
Step 3: Compute the sum of each x-value multiplied by its corresponding y-value
Step 4: Calculate the slope of the line using the formula:

$$m = \frac{\sum xy - \frac{(\sum x)(\sum y)}{n}}{\sum x^2 - \frac{(\sum x)^2}{n}}$$

where, x and y are the coordinates of the black pixels, n is the total number of black pixels and $m = \tan(\theta)$ (slope of the textual component). From the slope, the skew angle is calculated using $\theta = \tan^{-1}(m)$ for the textual line.

Store all the calculated skew angles satisfying $-30 < $ skew angle $< +30$ in the list £(t).

F. Calculation of final skew angle

In this step, all the elements of £(t) are sorted in increasing order and stored in £sorted(t). The final skew angle calculation for the document is as mentioned below.

$$\text{Skew of the image} = \left(\sum_{i=\eta_1}^{\eta_2} £_{sorted}(t) \right) \Big/ |\eta_2 - \eta_1 + 1|$$

where, $\eta_1 = (0.4 * \lambda)$; $\eta_2 = (0.6 * \lambda)$.

3   Experimental Results

The proposed approach has been designed and implemented in C language on an Intel core 2 duo processor (2.20 GHz) with 2 GB RAM. The test database constituted of 20400 different images having various skew angles. The images have been captured from different sources like journals, textbooks, newspapers etc. The content of the images varies from printed text (English and Hindi) to Pictures and stamps.

The Table 1 above shows the accuracy comparison of the proposed approach to the approach using horizontal projection of data for skew calculation [1]. It is evident from the results that the deviation of the proposed approach from the exact skew angle is lesser when compared to the other approach. Thus, using the angle obtained from the proposed approach for deskewing an image will result in better extraction accuracy as compared to that of other approach.

Table 1. Accuracy comparison of proposed approach with Horizontal Projection Profile method

Skew estimation techniques	Error deviation (degree) (Range −15 to +15)	Error deviation (degree) (Range −25 to +25)
Horizontal projection profile [2]	0.26	0.45
Proposed approach	0.16	0.33

Table 2 on the other hand shows the performance comparison between proposed approach and the horizontal projection profile method [1]. The comparison has been shown for images of A4 dimensions and scanned at 200 and 600 DPI. The timings of the proposed approach on both the type of images is better when compared to the other approach. This way the claim that the proposed approach is faster in comparison to the projection based approaches stands valid.

Table 2. Performance comparison of proposed approach with Horizontal Projection Profile method

		Time(in ms)
A4 size at 200 DPI	Horizontal projection profile [2]	**136.63**
	Proposed approach	**16.72**
A4 size at 600 DPI	Horizontal projection profile [2]	**274.58**
	Proposed approach	**89.35**

4   Conclusion

Skew detection plays a vital role in data extraction from document images using various ICR and OCR engines. In order to achieve high accuracy the extraction engines require that the data present in the image be free from skew. But with the increasing number of documents and the use of ADFs for scanning, has increased the risk of skew insertion in documents. As proven through the results on the 20400 test images, the performance and accuracy of the proposed approach is better than the projection based approaches. Future extension to the approach will handle skew angles above the ±30° range of the proposed approach. Also, the future extension will be able to support all orientations.

Acknowledgment. The authors would like to thank the reviewers for their valuable time and comments that made this paper in presentable form.

References

1. Li, S., Shen, Q., Sun, J.: Skew detection using wavelet decomposition and projection profile analysis. Pattern Recogn. Lett. **28**, 555–562 (2007)
2. Bagdanov, A., Kanai, J.: Projection profile based skew estimation algorithm for, JPEG compressed images. In: Proceedings of the 4th International Conference on Document Analysis and Recognition, Ulm, Germany, pp. 401–405, August 1997
3. Amin, A., Fisher, S., Parkinson, A., Shiu, R.: Comparative study of skew detection algorithms. J. Electron. Imag. **5**, 443–451 (1996)

4. Otsu, N.: A threshold selection method from gray-level histograms. IEEE Trans. Syst. Man Cybern. **9**(1), 62–66 (1979)
5. Dillencourt, M.B., Samet, H., Tammininen, M.: General approach to connected-component labeling for arbitrary image representations. J. ACM **39**(2), 253–280 (1992)
6. Srihari, S.N., Govindaraju, V.: Analysis of textual images using the Hough transform. Mach. Vis. Appl. **2**(3), 141–153 (1989)
7. Hinds, S., Fisher, J., D'Amato, D.: A document skew detection method using run-length encoding and the Hough transform. In: Proceedings of 10th International Conference on Pattern Recognition, pp. 464–468 (1990)
8. Flecher, L.A., Kasturi, R.: A robust algorithm for text string separation from mixed text/ graphics images. IEEE Trans. Pattern Anal. Mach. Intell. **10**, 910–918 (1988)
9. Cao, Y., Wang, S., Li, H.: Skew detection and correction in documents images based on straight-line fitting. Pattern Recognit. Lett. PRL **24**(12), 1871–1879 (2003)
10. Soille, P., Breen, E., Jones, R.: Recursive implementation of erosions and dilations along discrete lines at arbitrary angles. IEEE Trans. PAMI **18**(5), 562–566 (1996)

Classless Logical Regularities
and Outliers Detection

Alexander Dokukin$^{(\boxtimes)}$

Federal Research Center "Computer Science and Control" of Russian Academy
of Sciences, 119333 Vavilova Str. 40, Moscow, Russian Federation
dalex@ccas.ru

Abstract. The paper defines classless logical regularities in a similar
way to widely known logical regularities of classes. The main reason of
their introduction is unsupervised outlier detection in the areas where
main part of available data represents normal situation and outliers are
rare and different from each other. The substantiation and formal definition is followed by the method for their search and a real world example.

Keywords: Logical regularity · Classless regularity · Outlier detection

1 Introduction

The basic definition of the logical regularities of classes is given in [1]. They are
used in supervised classification and correspond to rectangular areas in feature
space that contain precedents of a single class.

Here is the formal definition of a logical regularity of a class. Let's consider a
recognition task and especially a training set $S = \{S_1, \ldots, S_m\}$. Each precedent
is a vector of n real values corresponding to object's features $S_i \in \Re^n$. The
set is divided into classes $K_1, \ldots, K_l$ (for simplicity classes don't overlap) and
classification of the training objects is known, i.e. $S_i \in K(S_i)$, $i = 1, \ldots, m$. Let's
consider also hyperrectangles of the following structure

$$R = r_1 \times \ldots \times r_n,$$
$$r_i = \begin{cases} [a_i, b_i], \\ (-\infty, b_i], \\ [a_i, \infty), \\ (-\infty, \infty), \end{cases} \tag{1}$$
$$i = 1, \ldots, n.$$

A hyperrectangle (1) is called a logical regularity of the class K_j if the following conditions hold:

$$\forall S \notin K_j, \ S \notin R, \tag{2}$$
$$|\{S | S \in K_j\}| \to \max. \tag{3}$$

© Springer International Publishing AG 2017
V.V. Krasnoproshin and S.V. Ablameyko (Eds.): PRIP 2016, CCIS 673, pp. 44–52, 2017.
DOI: 10.1007/978-3-319-54220-1_5

A generalization of the logical regularity that allows entry of other classes' objects to it is called a partial logical regularity. First, a fixed part of such objects was allowed [1]. Later a more generalized version was described that involves information value of the regularity. There are different ways of defining the information value which is a functional that reward presence of the target class precedents and penalize presence of the rest of them in it [2]. The most important of them for the purposes of the present article is statistical information value:

$$I_h(R, S) = -\ln h \left(\begin{matrix} p_j(R) & n_j(R) \\ P_j & N_j \end{matrix} \right),\tag{4}$$

where $h \left(\begin{matrix} p & n \\ P & N \end{matrix} \right)$ is hypergeometric distribution $h \left(\begin{matrix} p & n \\ P & N \end{matrix} \right) = \frac{C_P^p C_N^n}{C_{P+N}^{p+n}}$, $p_j(R)$– number of class K_j objects in R, P_j–number of all class K_j objects, $n_j(R)$– number of objects of other classes in R and N_j–number of all objects of other classes. The greater the $I_h(R, S)$ the lesser the probability to choose such a combination of objects from the set. In addition $p_j(R)$ is required to be greater than $n_j(R)$.

Thus, a generalized regularity is defined as an area in feature space with great enough information value. Rectangular shape is not necessary but it has certain advantages.

Logical regularities based classification is made by voting over the number of regularities of different classes covering a specific object. At that regularities can be weighed according to their quality as a part of class' objects covered by them or in general case according to their information value. A number of other conclusions beside object classification can be made from a set of logical regularities. They are used to acquire simple class descriptions in human readable form, to assess importance of a feature or typicality of the precedent [3]. The latter can be used to detect outliers of a specific class.

The basic idea behind mentioned applications of logical regularities of classes is very simple and it works well in practical tasks [3]. The quality of an entity is assessed as a weighed part of logical regularities set that involves the entity. However, there is one major drawback of the described methodology it requiring information about classification of the studied objects. It is hard to construct nontrivial regularities in a case where most precedents represent normal situations whereas outliers are few and very different. For example an analysis of surgery outcome in a good clinic can be considered [4], where complications are scarce and diverse. Moreover, a situation is feasible in which even the described scarce information is unavailable. For example, in the mentioned medical tasks an outcome that is considered generally well can require more attention. Thus, the analysis requires unmarked data.

To overcome the limitation while preserving the rest of the developments in the area the concept of classless regularities is presented. The formal definition will be given in the following section but the idea is similar to the logical regularities of classes though instead of penalty for the inclusion of objects of other classes the size of the area is penalized.

2   Classless Regularities

Classless regularity is an area of feature space defined by a specific information value functional.

Let's consider a training set of precedents $S = \{S_1, \ldots, S_m\}$ of the same structure as in the previous section ($S_i \in \Re^n$), but without any classification information. Let's also define the containing hyperrectangle R_S which is the minimal hyperrectangle containing S. The statistical information value of an arbitrary hyperrectangle $R \subset R_S$ is defined as follows:

$$I_b(R, S) = -\ln b\left(k(R), m, \frac{V(R)}{V(R_S)}\right),\tag{5}$$

where $b(k, n, q)$ is binomial distribution $b(k, n, q) = C_n^k q^k (1-q)^{(n-k)}$, $k(R)$–number of objects in R, $V(R)$–volume of $R \cap R_S$. Again, the greater the $I_b(R, S)$ the lesser the probability of such a number of uniformly distributed in R_S objects getting into the hyperrectangle. But to prevent rectangles being chosen for the rarity of their complement an additional condition is checked:

$$q^k < (1-q)^{(n-k)}.\tag{6}$$

The functional (5) with the condition (6) has all the required properties to serve as an information value. It increases with the number of objects getting into the regularity and decreases with its volume. It is also important that the functional doesn't depend on the scale of the features, because only the volume ratio is involved in the formula. Thus, we define the classless logical regularity as a hyperrectangle with great enough information value $I_b(R, S)$.

To illustrate the concept let's consider a normally distributed two-dimensional set of 100 dots and a set of 100 classless regularities imposed over it (see Fig. 1). The regularities for the example are obtained via the random search.

The intensity of gray corresponds to the number of regularities overlapping the specific area and subsequently the estimated typicality of its precedents. Though the rectangular shape of regularities infer some limitations to the picture the general idea is quite clear. The central area and neighboring dense extensions are considered most typical and the outbound singular objects are estimated outliers.

The limitations will be more obvious on the subsequent figures. The two-cluster structure on the Fig. 2 makes the in-between area quite typical, though clusters are formed by the similar distribution as on Fig. 1. But if considered separately the same clusters will produce a different picture more resembling the Fig. 1.

Quite the opposite can be said about Fig. 3. Though the set of dots is circular and quite symmetric, the regularities tend to horizontal and vertical edges making the resulting picture more square like.

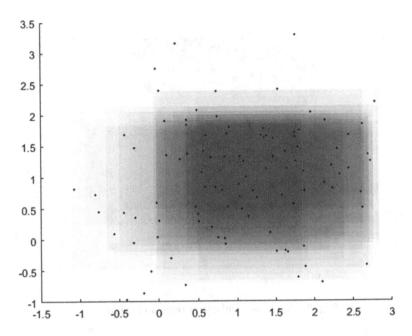

Fig. 1. Classless regularities over a set of random dots

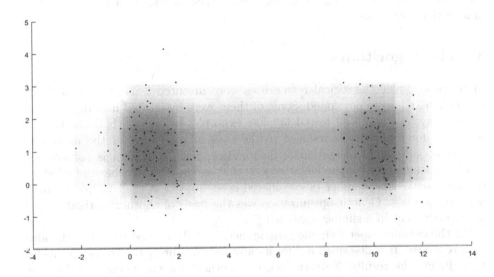

Fig. 2. Classless regularities over a two-cluster set

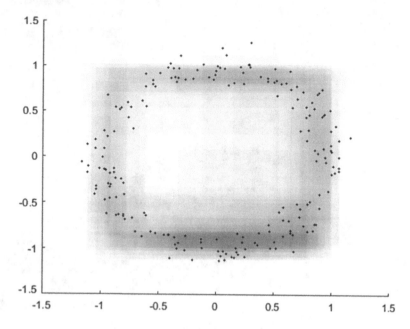

Fig. 3. Classless regularities over a circular set

Everything said it can be mentioned though that the same limitations stand for logical regularities of classes as well. And despite the fact the whole concept works quite well. Thus, the next section will be devoted to the description of the actual algorithm for searching the classless regularities which hopefully will inherit that performance.

3   The Algorithm

In the years since the logical regularities were invented a number of methods for their search were developed. Some of them were integrated into Recognition software system and well tested [3]. For example, in one method the task of searching for regularity is reduced to the task of searching for the maximum consistent subsystem of system of inequalities. Such a system is constructed in correspondence to a particular object that is supposed to be contained by the regularity. The number of considered seed objects is then estimated by a permutation test. Genetic optimization was the basis of another method as well as fastest ascent in a similar approach [5].

In the current paper a simple genetic method will be used to justify the idea with real data. It is described not for its novelty but purely to provide means for reproducing the results. Nevertheless, the method has one peculiarity because it's aimed for searching a diverse set of regularities in different parts of the considered space instead of one optimal solution and its neighbors. At that the two modifications will be considered. The first one is a straight forward genetic search for the set of best regularities. The second one will inherit the concept of a seed object. Thus, it will search for a set of best regularities with respect to a single precedent and then the found sets will be combined.

So, the training sample $S = \{S_1, \ldots, S_m\}$ is considered, each precedent being a real vector $S_i = (a_{i1}, \ldots, a_{in}) \in \Re^n$. A grid of possible regularities' boundaries is calculated then. Let $\{a_{1j}, \ldots, a_{m_j j}\}$ be the set of unique values of j-th feature, $2 \leq m_j \leq m$. The respective set of boundaries will contain the following values:

$$
\begin{aligned}
b_{1j} &= a_{1j} - \Delta_j, \\
b_{ij} &= \frac{a_{(i-1)j} + a_{ij}}{2}, \quad i = 2, \ldots, m_j, \\
b_{(m_j+1)j} &= a_{m_j j} + \Delta_j.
\end{aligned}
\tag{7}
$$

Here Δ_j is half of an average distance between the sample values. When describing the regularity informally the boundary values can be treated as infinity.

The initial population P_0 is defined by selecting left and right boundaries randomly from the respective sets:

$$
\begin{aligned}
R_i &= r_{i1} \times \ldots \times r_{in}, \quad i = 1, \ldots, N_0, \\
r_{ij} &= [b_{q_i j}, b_{t_i j}], \quad q_i, t_i \in 1, \ldots, m_j + 1, \quad q_i < t_i,
\end{aligned}
\tag{8}
$$

where N_0 is population size.

Each next generation P_k, $k > 0$ is obtained in following steps.

Step 1. A set of c randomly selected pairs of regularities from P_{k-1} is used for producing a posterity C_k. For each pair R_x, R_y a random value $p_{xy} \in \{1, \ldots, n-1\}$ is generated and C_k is supplemented by a new regularity $R_{xy} = r_{x1} \times \ldots \times r_{x p_{xy}} \times r_{y(p_{xy}+1)} \times \ldots \times r_{yn}$.

Step 2. Objects of P_{k-1} undergo mutations to form the set M_k. It means that each interval of each regularity with some probability μ is replaces with a random one.

Step 3. The whole set $P_{k-1} \cup C_k \cup M_k$ is cleared from duplicates and N_0 of its best objects are transferred to the P_k generation.

Thus, the first algorithm has four parameters: N_0, c, μ, and the number of generations produced g. The final generation P_g is denoted as P.

The second algorithm is mostly the same but the initial random population and mutations are performed in respect to a certain object $S_\nu \in S$. That is the formulae (8) is updated with condition $q_i < a_{\nu j} < t_i$. The parameters are supplemented with s which is a number of seed objects to use. The resulting algorithm is executed multiple times in respect to each of s randomly selected objects and $\sim N_0/s$ part of each resulting set is extracted to form the resulting set P of N_0 regularities.

The difference between the algorithms can be clearly seen in the following two images (see Figs. 4 and 5). The algorithms are applied to the same random sets of dots with the same set of parameters $N_0 = 1000$, $c = 1000$, $\mu = 0.01$, $g = 10$. At that 20 random seeds are used in the second case and 200 best regularities shown.

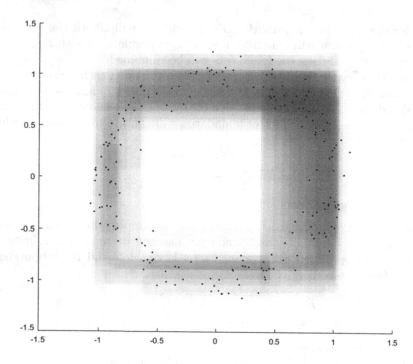

Fig. 4. Results of the first algorithm

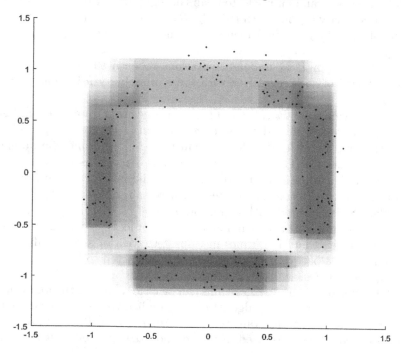

Fig. 5. Results of the second algorithm with 20 seeds

As can be seen from the figures the second algorithm provides more uniform coverage of dots that better corresponds to expectations. Thus, the second one will be used to analyze data in the following section.

4   Real World Tasks

Despite artificial examples shown above the proposed method for outlier detection has its drawbacks. The earlier mentioned shape problems are one of them but there are more. For example, missing values or categorical features need special approach. So, a single set of neurological data was chosen to illustrate method's potential, for it has no missing values and all features are binary or quantitative. It is data about outcomes of ischemic stroke [4]. 125 patients are described by 30 features including age, gender, risk factors, anamnesis and so on. Outcomes of disease are also known, bad ones are much rarer as 95 patients survived.

The data was used for unsupervised outliers detection with the proposed method. At that the second modification of the algorithm was applied with the following parameters: $N_0 = 100$, $c = 10000$, $\mu = 0.1$, $g = 10$ and 20 seeds. The resulting numbers of covering regularities were used to estimate patients' typicality $T(S)$ and compared to the known outcome of the disease, i.e.

$$T(S) = \frac{|\{R \in P \mid S \in R\}|}{N_0}. \tag{9}$$

Correlation coefficient between the estimate and the outcome was 0.3743 which is significant at $p = 2 \cdot 10^{-5}$.

5   Conclusion

The concept of classless logical regularities has been proposed. Those are rectangular areas of feature space with statistically justified high density of precedents. For that a special information value functional has been suggested which rewards presence of precedents and penalizes size of the regularity.

The goal of the proposed concept is unsupervised outlier detection that is aimed at finding small number of diverse exceptions in a mostly "normal" data. An algorithm has been proposed to illustrate its potential using artificial and real world tasks. At that simple artificial tasks generate plausible picture and the real one shows significant correlation between the calculated estimate and a real hidden parameter.

The algorithm has its drawbacks though. Namely, it deals poorly with missing values and categorical features. But this is only the first attempt and further research will be aimed at removing those limitations. It is also important to try other approaches beside genetic optimization and to test other applications of logical regularities such as feature selection.

Acknowledgments. The work was supported by Russian Foundation for Basic Research, grants 15-07-00980 and 17-01-00634.

References

1. Ryazanov, V.: Logical regularities in pattern recognition (parametric approach). Comput. Math. Math. Phys. **47**(10), 1720–1735 (2007). doi:10.1134/S0965542507100120
2. Vorontsov, K.: Lekcii po logicheskim algoritmam klassifikacii [Lections on logical classification algorithms], Moscow (2007) (in Russian). http://www.ccas.ru/voron/download/LogicAlgs.pdf
3. Zhuravlev, Y., Ryazanov, V., Senko, O.: RASPOZNAVANIE. Matematicheskie metody. Programmnaja sistema. Prakticheskie primenenija. [RECOGNITION. Mathematical Methods. Software System. Practical Solutions.], Phasis, Moscow (in Russian) (2006)
4. Zhuravlev, Y., Nazarenko, G., Vinogradov, A., Dokukin, A., Katerinochkina, N., Kleimenova, E., Konstantinova, M., Ryazanov, V., Senko, O., Cherkashov, A.: Methods for discrete analysis of medical data based on recognition theory and some of their applications. Pattern Recog. Image Anal. Adv. Math. Theo. Appl. **26**(3), 643–664 (2016). doi:10.1134/S105466181603024X
5. Dokukin, A.: Use of information value in avo-polynomial method training. Inf. Models & Analyses **2**(2), 123–126 (2013)

Ensembles of Neural Network for Telemetry Multivariate Time Series Forecasting

Alexander Doudkin and Yauheni Marushko[(✉)]

United Institute of Informatics Problems of National Academy of Sciences of Belarus,
st. Surganova 6, Minsk, Belarus
doudkin@newman.bas-net.by, marushkoee@gmail.com

Abstract. In this paper, we propose to solve the problem of forecasting multivariate time series of telemetry data using neural network ensembles. Approaches to the forming neural network ensembles are analyzed and prediction accuracy is evaluated. The possibility of training the neural network ensembles is studied for reducing errors of multivariate time series forecasting.

Keywords: Forecasting · Artificial neural network · Ensemble of neural networks · Telemetry · Multivariate time series

1 Introduction

Space telemetry is a set of technologies that allows remote collection of information about on-board spacecraft subsystems. The subsystems are controlled by analysis of sensor readings that are distributed across submodules. A subsystem state at a particular point in time is described by a vector of sensor values. The time sequence of states is a sequence of vectors of sensor values. Hence, space telemetry data are multivariate time series. One of analysis tasks is the forecasting of such time series.

The task of forecasting multivariate time series is generally formulated as follows [1, 2]: from the known current value of the sequence $y(k)$ and the prehistory $y(k-1), y(k-2), \ldots, y(k-m)$ we should evaluate the next value $\hat{y}(k+1)$. Each element of the sequence $y(k)$ represents a vector of values at time k. The length sequence m is called as time window.

A variety of techniques has been used in short-term forecasting, including regression and time series analysis. Simple regression and multiple linear regressions are frequently used. They have an advantage that they are relatively easy for implementation. However, they are somewhat limited in their ability to forecast in certain situations, especially in the presence of nonlinear relationships between high-level noisy data. Most time series models also belong to the class of linear time series forecasting, because they postulate a linear dependency between the value and its past value. The autoregressive moving average ARMA model and its derivatives are often used for the case of univariate analysis However, the artificial neural networks (NNs) often outperform these models in solving complicated tasks [1]. Deep neural networks can also be used [3], but the large training set is needed in this case.

© Springer International Publishing AG 2017
V.V. Krasnoproshin and S.V. Ablameyko (Eds.): PRIP 2016, CCIS 673, pp. 53–62, 2017.
DOI: 10.1007/978-3-319-54220-1_6

The processing and analysis of the telemetry data is accompanied by non-deterministic noises. In this case it is preferred to use NNs technology. The effectiveness of this technology depends on NN architectures and learning methods [1, 4], which requires multiple experiments.

There are examples of using NNs in on-board intelligent decision support systems for managing complex dynamic objects and diagnosis of its condition [5, 6].

In this paper, we investigate the possibility of short-term prediction parameters of telemetry using a neural network ensemble (ENN) [7, 8], which is a set of NNs and makes decisions by averaging the results of the separate NNs.

Predictive analytics and machine learning are often faced with the "concept drift", which means that the statistical properties of the target variable, which the model tries to predict, change over time in an unpredictable manner [9] and this increases the prediction error. Hence, the neural network prediction efficiency can be improved by using iterative learning methods [7, 9, 10]. These methods involve accuracy estimation of the models and their ranking on each analysis iteration. In the case of lowering the overall accuracy, the ensemble detects the concept drift and a new NN, trained on the relevant data, is added to the ensemble. In this approach, the model laid during initial training is retained and the new parameters are entered without "forgetting" problem. Thus, an additional learning ENN is realized.

2   Methods and Algorithms

The main objective of this study is to explore the possibility of using ENNs for telemetry data forecasting. Therefore, NNs and ENNs are selected as the test models. Classic feed forward NN with one hidden layer is selected as a single NN model.

A comparative analysis of the following approaches to the formation of the output value of the ensemble is performed.

1. The output value is formed as the sum of the individual networks outputs. It is calculated for the case with a single output neuron by the formula:

$$y = \frac{1}{n} \sum_{i=1}^{n} y_i, \tag{1}$$

where n – the number of the networks in the ensemble, y_i – the output of i-th network;
2. The output value is formed as a weighted sum of the individual networks outputs. It is calculated for the case with a single output neuron by the formula:

$$y = \sum_{i=1}^{n} y_i \cdot w_i, \tag{2}$$

where n – the number of the networks in the ensemble, y_i – the output of i-th network, w_i – the weight of i-th network, which is formed according to the formula:

$$w_i = \frac{mse_i}{\sum\limits_{i=1}^{n} mse_i},$$

(3)

where mse_i – MSE of i-th network on a validation set;

3. The output value is formed as the weighted sum of the outputs of the individual networks (formula (1–3)) and weighing is repeated after a certain interval of time samples with the evaluation on this interval (dynamically weighted ensemble).

Also an iterative method of ensemble learning was studied. The following algorithm was used for this purpose:

1. Processing of the current input vector.
2. Evaluation of the accuracy of the ensemble. For this purpose, errors in the previous step and the current are compared.
3. If the error is not increased or increased in a predetermined range, go to the next input vector.
4. Otherwise, the training set is formed, which includes all the accumulated data from the last additional training.
5. The formation and training of a new neural network.
6. Formed network is added to the ensemble.
7. Recalculation of the weighting coefficients based on their errors on the latest data produced for all neural networks of the ensemble.

3   Experiment

The data set is a finite set of precedents, which is selected in some way from the set of all possible precedents called the general population.

Data sets parameters for our experiments are presented in Table 1. Each set of telemetry data is obtained by sensors of the correction propulsion system. These include temperature parameters and pressure levels of the xenon supply unit, electric parameters of flow control, electrical parameters of the engines anode and cathode. Values provided by the sensors depend on the mode which is set by the control commands. Sensor values are correlated with each other; it allows expecting a satisfactory assessment of the forecast.

Table 1. Telemetry data of the correction propulsion system of the spacecraft

Title	Sampling time, s.	The dimension of the time series	Number of samples
Dt_set_s01	1	24	57501
Dt_set_s05	0.5	24	12245
Dt_set_s1	0.1	24	6613

Since the learning is carried out with the teacher, it is necessary to form the learning set of pairs "input vector, output vector". Formation of the pair of the learning sample is carried out by windowing method [2]: it takes a certain period of time series and a few observations stand out from it, and that will be the input vector. The value of the desired output in the training example will be the next in order. Then, the window moves in the one position in the direction of increasing time, and the process of forming the next couple of training sample is repeated. Thus, if the dimension of the time series data is N and the window size — W, the neural network should receive the input sample with size $N \times W$. So, for the window $W = 20$ Dt_set_s1 set is converted into an input set with size of 6589×480, and target set — with size 28×6589.

Resampling and scaling performed during the preparation of the datasets of telemetric sensor information.

Resampling is performed to convert the raw data, representing as a sequence of time stamps of important events to the form with fixed sampling time. Scaling is necessary to bring the data in the valid range $[-1, 1]$. Also, the outputs of the network are scaled.

The aim of the experiment is to determine an influence of the parameters of a single neural network and ensembles on forecast performance.

Multilayer perceptron with one hidden layer with *hyperbolic tangent* non-linear activation function is used as the base element of the ensemble. RPROP algorithm is used to train a single network [11]. Prediction window is chosen to be 20 samples.

To evaluate the quality of trained neural networks as well as to compare different ensembles, the following values are used:

- mean square error, MSE:

$$\text{MSE} = \frac{1}{m} \sum_{i=1}^{m} e_i^2; \tag{4}$$

- mean absolute error, MAE:

$$\text{MAE} = \frac{1}{m} \sum_{i=1}^{m} |e_i|. \tag{5}$$

where $e_i = y_i - t_i$, y_i and t_i – the obtained and the desired output signals of i-th neuron of the output layer, respectively, m - the size of the output layer of the neural network.

In this experiment, the input set is divided in a ratio of 9:1 for a general training set and the final test set. The general training set is divided into validation set (15%), a test set (15%) and training set (70%) randomly, which were used for training, evaluation and search for the best architecture, respectively.

The resulting test set is used to calculate the final estimates obtained by neural networks.

3.1 Evaluation of the Hidden Layer Size

Suboptimal size of the hidden layer of the neural network was evaluated according to the following algorithm.

1. Determination of the search interval.
2. Training of 10 networks with the current hidden layer size, which is selected from the search interval.

3. Formation of the weighted ensemble.
4. Evaluation of the ensemble accuracy.
5. Until the end of the search range, go to the next element of the interval (step 2).
6. Select the ensemble with the lowest MSE on search range, the size of the hidden layer of the ensemble element is the best one.

The dependence of the accuracy of the hidden layer size is shown in Fig. 1.

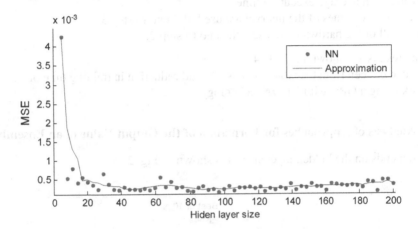

Fig. 1. Evaluation of the hidden layer size for dataset Dt_set_s1

The sizes of the hidden layers for the data sets are shown in the Table 2.

Table 2. Evaluation of the hidden layer size

Dataset	Hidden layer size
Dt_set_s1	28
Dt_set_s05	33
Dt_set_s01	70

It should be noted that this algorithm significantly increases the formation of the ensemble.

3.2   Evaluation of Neural Network Training Time

The aim of the experiment is to determine the differences in the speed of execution of training the neural network on a single central processing unit (CPU) and graphics processing unit (GPU).

Hardware for the experiment:
CPU Intel Core i5 4200H, 2 cores, 4 threads, 2 800 MHz (Turbo 3 400 MHz); GPU NVIDIA GeForce GTX 860 M, 1029 МГц, 640 CUDA cores.
hardware were chosen based on the close price.

MATLAB is chosen as an experimental platform. Hidden layer size varied in the range [4, 56] neurons.

1. The evaluation procedure includes the following steps:
2. Select the hardware (CPU, GPU).
3. Select a value from the range of the hidden layer.
4. Train 10 single neural networks.
5. Evaluate an average execution time.
6. If not all the values of the interval are used, then go to step 3.
7. If not all of the hardware are used, then go to step 2.

The results are shown in Table 4.

As can be seen from the table, there is a 5-fold reduction in training time of a neural network using a GPU with CUDA technology.

3.3   Analysis of Approaches for Formation of the Output Value of an Ensemble

MSE depends on the hidden layer size (it is shown in Fig. 2).

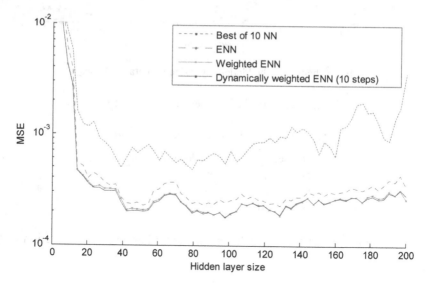

Fig. 2. MSE of different models depending on the size of the hidden layer

Evaluation of different models across testset Dt_set_s1 is given in Table 3.

Table 3 shows that dynamically weighed ensemble has the smallest error and the difference in the estimated parameters for weighted ensembles is very small.

3.4   Evaluation of Dynamically Weighted of Ensemble

Evaluation of weighting interval for dynamically weighed of the ensemble is performed. MSE plot for the various parameters is shown in Fig. 3.

Table 3. Evaluation of training time

Hidden layer size	Matlab CPU, s.	Matlab GPU (CUDA), s.	Speed up
4	93,9962	17,2422	5,4515
12	109,2202	19,5071	5,5990
20	86,2531	18,4156	4,6837
28	105,0203	20,0411	5,2402
36	113,4552	18,902	6,0023
44	90,6871	19,8095	4,5780
52	95,4515	16,4389	5,8064
56	107,1488	20,8944	5,1281
Mean			5,3112
Median			5,3459
Minimum			4,5780

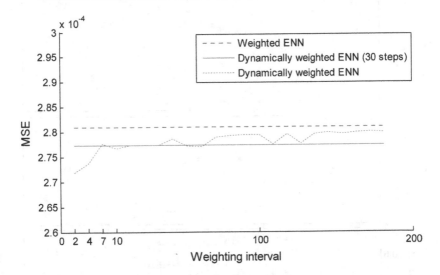

Fig. 3. Evaluation of weighting interval for the dynamically weighted ensemble

The evaluation procedure includes the following steps:

1. Training the ensemble with the sub-optimal size of the hidden layer.
2. Determination of the search interval of weighting steps.
3. Evaluation of MSE for different models and the ensemble with the current step of weighting repetition.
4. Until the end of the search interval, go to the next element of the interval.

Figure 3 shows that the dynamic weighting ensemble with a small step of repetitive weighting (less than 10 samples) has the smallest error.

3.5   Evaluation in the Case of Concept Drift

Concept drift refers to a change in value over time and, consequently, to a change in the distribution of the value. The environment from which these values are obtained is not stationary.

An iteratively trained (without previous access to the data) expert ensemble in combination with some form of weighted voting for the final solutions is used in drift detection algorithms [9, 10].

For this experiment, modifications have been made artificially in the investigated data. The linearly increasing trend was added, and the sine wave signal was modeled as a periodic component.

Modification of Dt_set_s1 dataset is shown in Fig. 4. The first of the 1319 accounts are used without modification.

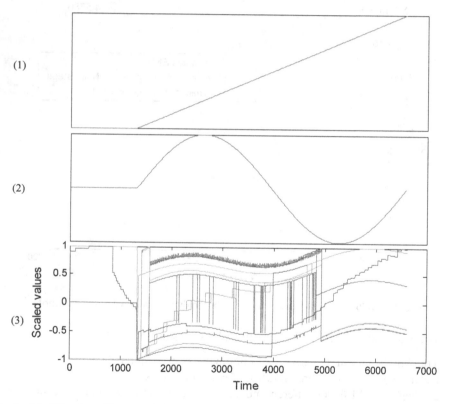

Fig. 4. Modification of Dt_set_s1 data set: (1) Linearly increasing trend; (2) periodic component; (3) a modified multivariate signal

The evaluation procedure includes the following steps.

1. Train the ensemble.
2. Modify the data set by adding the trend and (or) the seasonal component.
3. Set the error threshold for additional training algorithm.

4. Set a minimum amount of data for accumulation.
5. Rate accuracy of different models, as well as an ensemble with additional training (additional training is performed only when the accumulated needed amount of data).

Evaluation of different architectures on a modified dataset are shown in Table 4.

Table 4. Evaluation of NN models for dataset Dt_set_s1

Model	MSE, 10^{-4}	MAE, 10^{-3}
Single neural network	3.51	0.115
Ensemble of neural networks	3.66	0.114
Weighted ensemble of neural networks	2.76	9.65
Dynamically weighted ensemble of neural networks, interval = 10	2.75	9.59

All NN models showed a significant drop in the accuracy on a modified set including the ensemble with additional training. This is associated with accumulation interval for the additional data set (Table 5).

Table 5. Evaluation of different architectures on a modified set

Model	MSE	MAE
Single neural network	0.6035	0.4843
Ensemble of neural networks	0.0304	0.1274
Weighted ensemble of neural networks	0.0289	0.1212
Dynamically weighted ensemble of neural networks, interval = 10	0.0266	0.1170
Dynamically weighted ensemble of neural networks, interval = 10 with additional training	0.0153	0.0877

4 Conclusion

The developed ENNs significantly reduce the forecasting error in comparison with single NNs.

Dynamically weighted ENN with a small step of repetitive weighing (less than 10 samples) has the smallest error and the difference in the estimated parameters for weighted ensembles is very small.

The minimum mean square error at short-term forecasting of telemetry data obtained by sensors of the correction propulsion system is equal to $2,75 \times 10^{-4}$.

All NN models showed a significant drop in the accuracy on a modified set, including the ensemble with additional training. The ensemble of neural networks with additional training also showed a drop in precision due to needing of additional network training on enough samples of new data, but the resulting accuracy was higher.

Testing also showed that NN training on a single GPU with CUDA technology gave 5-fold reduction in the time in comparison with the CPU in the specified equipment.

References

1. Quan, H., Srinivasan, D., Khosravi, A.: Short-term load and wind power forecasting using neural network-based prediction intervals. IEEE Trans. Neural Netw. Learn. Syst. **25**(2), 303–315 (2013)
2. Shumway, R.H., Stoffer, D.S.: Time Series Analysis and Its Applications. Springer, New York (2011). doi:10.1007/978-1-4419-7865-3
3. Dalto, M., Matusko, J., Vasak, M.: Deep neural networks for time series prediction with applications in ultra-short-term wind forecasting. In: IEEE International Industrial Technology, Seville, Spain, 17–19 March (2015)
4. Valipour, M., Banihabib, M.E., Reza Behbahani, S.M.: Comparison of the ARMA, ARIMA, and the autoregressive artificial neural network models in forecasting the monthly inflow of Dez dam reservoir. J. Hydrol. **476**, 433–441 (2013)
5. Khachumov, V.M., Talalaev, A.A., Fralenko, V.P.: Review of Standards and the concept of monitoring, control and diagnostics of the spacecraft tools building. Softw. Syst. Theor. Appl. **3**(26), 21–43 (2015). (in Russian), Volume 6
6. Emelyanov, Yu.G, Konstantinov, K.A., Pogodin, S.V.: Neural orientation angles and distance of the spacecraft sensor control system. Softw. Syst. Theor. Appl. **1**(1), 45–59 (2010). (in Russian)
7. Marushko, Y.: Using ensembles of neural networks with different scales of input data for the analysis of telemetry data. In: Proceedings of the XV Internship Ph.D. Workshop OWD 2013, Wisla, 19–22 October 2013, pp. 386–391 (2013)
8. Kourentzes, N., Barrow, D.K., Crone, S.F.: Neural network ensemble operators for time series forecasting. Expert Syst. Appl. **41**(9), 4235–4244 (2014). ISSN 0957-4174
9. Elwell, R., Polikar, R.: Incremental learning of variable rate concept drift. In: Benediktsson, J.A., Kittler, J., Roli, F. (eds.) MCS 2009. LNCS, vol. 5519, pp. 142–151. Springer, Heidelberg (2009). doi:10.1007/978-3-642-02326-2_15
10. Parikh, D., Polikar, R.: An ensemble-based incremental learning approach to data. IEEE Trans. Syst. Man Cybern. Part B Cybern. **37**(2), 437–450 (2007)
11. Riedmiller, M.A, Braun, H.: Direct adaptive method for faster backpropagation learning: the RPROP algorithm. In: Proceedings of the IEEE International Conference on Neural Networks (ICNN), San Francisco, pp. 586–591 (1993)

Comparative Analysis of Some Dynamic Range Reduction Methods for SAR Image Visualization

Valery Starovoitov[✉]

United Institute of Informatics Problems, Minsk, Belarus
valerys@newman.bas-net.by

Abstract. The main objective of the paper is to present our comparative investigations of high dynamic range compression to demonstrate the SAR images on monitors with the standard dynamic range. To display the images we need to compress the input image range in 256 times with preservation of the most object details and maximal image contrast. We studied several well-known tone mapping methods developed for optical images and used some published no-reference quality measures for evaluation of the obtained dynamic range compression results.

Keywords: SAR · HDR · Satellite imagery · Image processing

1 Introduction

Synthetic aperture radar (SAR) is a popular type of radars, which is used on satellites to create images of the Earth. SAR uses the motion of the radar antenna over a targeted region to provide finer spatial resolution than is possible with conventional beam-scanning radars. Modern satellite radar systems provide resolution of about 1 m in x-band with 2.4–3.75 cm wavelength. Signals of backscattered microwave pulses are reflected from the ground and storied in SAR images as pixels intensities. Pixel values are lower over flat ground. However, they increase in some cases due to backscattering from metal and stone objects. That is why the pixels have a high dynamic range of luminance levels: from 0 to 65535. SAR images belong to the class of High Dynamic Range (HDR) images.

The conventional method used to display SAR images on standard monitors contains a drawback: without compression of intensity with dynamic range, SAR images cannot be displayed with adequate contrast of all details. Standard or low dynamic range of monitors is [0–255] and may be abbreviated as SDR or LDR.

There is a very important problem to compress a HDR SAR image into an LDR-representation with minimal information lost and maximal contrast preservation. We denote this transformation as HDR-LDR transform.

The linear compression does not solve the problem of HDR-LDR transformation because from 80 to 90% of pixels have intensity values in the range [0–255], but large changes in pixel values occur around regions containing artificially made objects. That is why compression in 256 times produces mostly black image with some tiny gray or white blobs.

© Springer International Publishing AG 2017
V.V. Krasnoproshin and S.V. Ablameyko (Eds.): PRIP 2016, CCIS 673, pp. 63–76, 2017.
DOI: 10.1007/978-3-319-54220-1_7

In the literature, similar transformations are called tone mapping [1–3] and mainly are applied to images registered by optical systems. An ideal HDR-LDR transformation needs to be done with the minimum loss of information and correlate with the human visual system. All the tone mapping transformations may be divided into two types: global and local. Global means that a intensity transformation function depends on the intensity value of the processing pixel independently of other. Local one means that the transformation function depends on intensity values of a neighborhood of the processing pixel. As the result, transformations of the last type take more time.

We have formulated several requirements for a HDR-LDR transformation function f:

(I) Total ordering. Scene-to-monitor mapping must be monotonic, i.e. if there are two pixels with gray values $A_1 < A_2$, after a HDR-LDR transformation f brightness of the pixels should satisfy to the inequality $f(A_1) \le f(A_2)$. This means that the functions f must be nondecreasing ones.

(II) Preservation maximum contrast. Higher contrast of the resulting image means better quality of the transformation result. If for two pixels with gray values A_1 and A_2, we have $contrast(A_1) < contrast(A_2)$, after the transformation it must be $contrast(f(A_1)) \le contrast(f(A_2))$, where $contrast(Y)$ means contrast abound pixel Y. To measure quality of the result, we need a quality function. Compare image quality, in particular image contrast, after several transformations, we can select the better transformation for convenient visual evaluation of the result.

(III) Fast calculation. The transformation must be fast, because real SAR images in our experiments were up to 27083×43750 pixels and there are very few local transformation, which satisfy to this requirement.

In this study, we have limited our considerations by the most popular tone mapping transformations developed for optical images and evaluated their applicability to SAR images.

The other sections of the paper are organized as follow: in Sect. 2, we briefly discuss HDR-LDR transformations used in the study; in Sect. 3, we outline some quality measures used for the transformation comparison; in Sect. 4, the experimental results are presented.

2 HDR-LDR Transformations

Very few papers are devoted to HDR-LDR transformation of SAR images. In [1] the authors apply k-means image clustering into 6 classes, different intensity compression for every class, and some additional pre- and post-processing of the result. Unfortunately, the authors used only visual comparison of the studied methods.

In [2] several dynamic range reduction techniques known from the tone mapping of optical images were analyzed regarding their applicability to SAR data.

Nonlinear luminosity correction is a popular image processing method that was applied for TV-sets and monitors since the middle of the last century. J.R. Little in [4] have described mainly concave functions like logarithmic and exponential. In [5] it was shown similarity between the logarithmic type of image processing model and the

Naka–Rushton model used for description of the human visual system (HVS). Similar model was used by Reinhard et al. in [6].

We examined global tone mapping transformations developed for optical images but suitable for SAR imaging. All of them must be concave and may be divided into 2 groups:

- monotonically increasing nonlinear functions (like exponential, logarithmic, μ-law [7]),
- S-shaped functions (like sigmoid).

Following to recommendations of paper [2] we have selected the most popular HDR-LDR transformations that may be used for SAR image visualization. In the semi-logarithmic scale, the luminosity transformation functions studied in this paper looks as presented in Fig. 1. In the figure the following legend is used: Ashikhmin global and local means two variants developed in [8], Reinhard is the photographic tone mapping from [6], Reinhard-Devlin is the photoreceptor model from [9], Drago is the logarithmic mapping from [10] and mu-law is described in [7]. We also tested direct logarithmic mapping, gamma-based mapping and Schlick uniform rational quantization mapping [11].

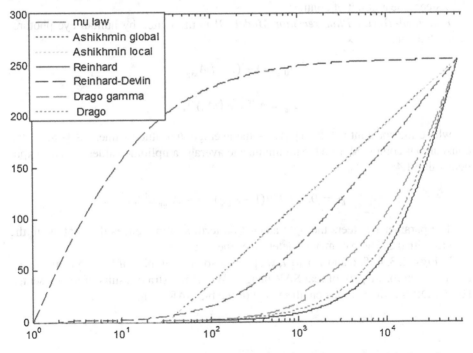

Fig. 1. Functions of several HDR-LDR transformations in the semi-logarithmic scale

Short descriptions of the mentioned measures are given below. In the following, the input data are amplitude values A. They are normalized to the range [0, 1], without normalization we denoted them as A_s. A_d means display luminance, $(L_d - 1)$ means display maximal gray value, $(L_s - 1)$ is the SAR image maximal gray value.

Direct logarithmic mapping may be described by the following function:

$$A_d = k^* \log(1 + cA_s) / \log(1 + c),$$ (1)

where c – is a constant,
k – is a normalizing coefficient.
Gamma Mapping:

$$A_d = (L_d - 1)^* (A)^g,$$ (2)

the exponent value $g > 0$ and it is the main parameter of the function.
Drago Logarithmic Mapping [10]:

$$A_d = [m / \log(1 + c)]^* [\log(1 + cA) / \log(2 + 8A)^b)]$$ (3)

where c – is analogous to the simple logarithmic mapping,
m determines the brightness of the result,
b steers the amount of contrast.
Reinhard–Devlin Photoreceptor Model [9] is motivated by human eye photoreceptor behavior.

$$A_0 = lA + (1 - l)A_{avg},$$

$$A_d = A / [A + (lA_0)^p],$$

where the constant $l \in [0, 1]$, A_{avg} – the average amplitude value, $l \in [-8, 8]$, the constant p is computed from the minimum and average amplitude values with the exponent value 1.4:

$$p = 0.3 + 0.7((1 - A_{avg}) / (1 - A_{min}))^{1.4}.$$

The parameter l steers the light adaptation term and influences the contrast in the resulting image. The parameter l determines the brightness.

In Figs. 2, 3, 4, 5, 6 and 7 below, we present some examples of high dynamic range compression for display of real SAR images. We demonstrate results of the presented HDR-LDR transformations on afragment of the big SAR image.

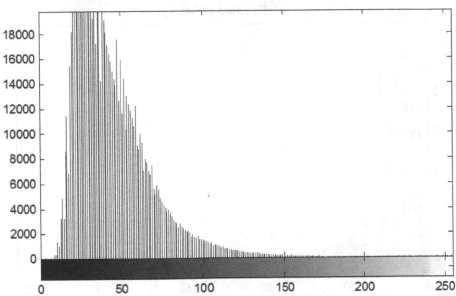

Fig. 2. A fragment of a SAR image compressed by mu-law transformation [7] and the image histogram

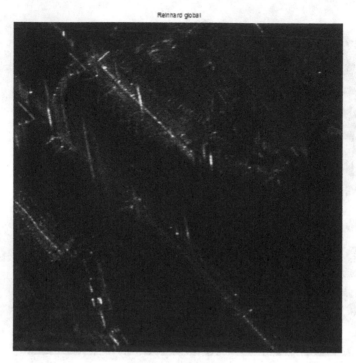

Reinhard global

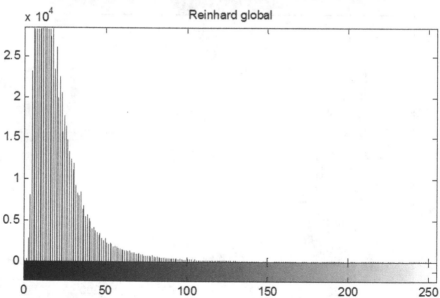

Fig. 3. The same fragment of a SAR image compressed by the Reinhard transformation [6] and the image histogram

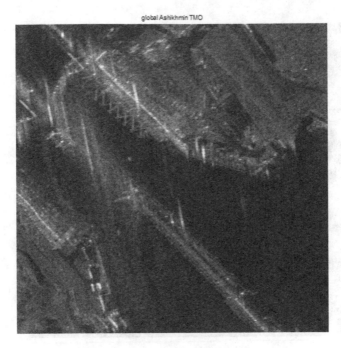

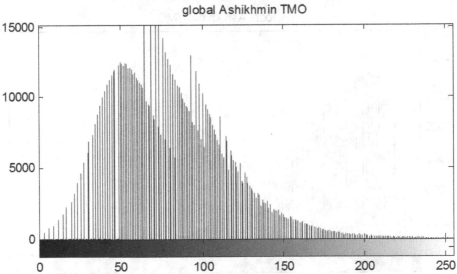

Fig. 4. The same fragment of a SAR image compressed by the global Ashikhmin transformation [8] and the image histogram

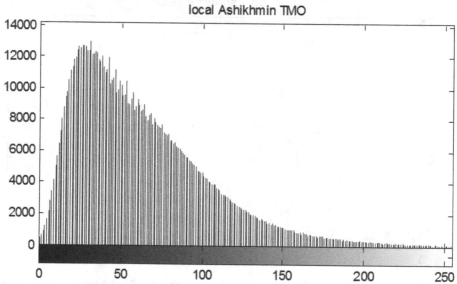

Fig. 5. The same fragment of a SAR image compressed by the local Ashikhmin transformation [8] and the image histogram

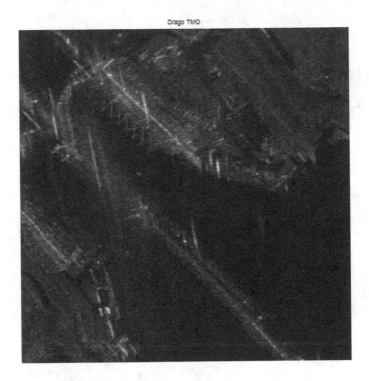

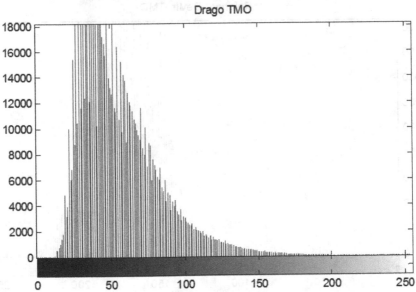

Fig. 6. The same fragment of a SAR image compressed by the Drago transformation [10] and the image histogram

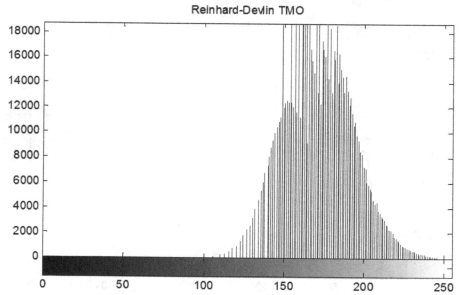

Fig. 7. The same fragment of a SAR image compressed by the Reinhard-Devlin transformation [9] and the image histogram

Visual quality evaluation of the presented figures gives us the following results. The background is usually dark, the metal and artificially created objects are very bright, speckle-noise and other kind of noise are also presented in the picture. The main objective of the tone mapping transformation is to get the best image contrast. Nevertheless, even for the six demonstrated pictures visually difficult to identify images with better contrast.

For several transformations mentioned in the previous section have infinite number of parameter variants. Because we reduce the original image, range in 256 times, two questions arises: (1) which transformation will produce better result, (2) which set of the transformation parameters will better present image contrast or sharpness. It is impossible to get answers using just visual evaluation of the results.

3 Quality Measures

To find answers one can use various image quality assessment (QA) measures. We do not have the ideal image for comparison with our transformed variant that is why we need in a so-called no-reference (NR) quality measure. Last decade no-reference measures were studied very actively, we just call a few major works such as [12–14].

The main features in our research are contrast, sharpness and naturalness of the image after dynamic range compression. The desirable NR-measure must be also highly correlated with the human-subjective evaluation of the image quality.

Among this class of measures, we have selected and studied TMQI metric [15] by Zhu and Milanfar, S3 measure of sharpness by Vu et al. [16], sharpness index by Blanchet and Moisan [17], sharpness metric by Leclaire and Moisan [18], BIQAA measure by Gabarda and Cristybal [19], the blur measure by Crete et al. [20]. We added to mentioned set of measures two sharpness measures calculated from gradients and image entropy. To save space, we will not describe these measures.

4 Experiments

We tested the mentioned tone mapping transformations and applied the mentioned NR image quality measures to the results obtained from real SAR images downloaded from the publicly available website [21].

We have formulated the following requirements to NR quality measure:

- the function should be fast,
- it smoothly changes the values when we smoothly change the HDR-LDR transformation parameters,
- it has one global extremum appropriate optimal set of the parameters,
- the extremal measure value correlate with the best human quality image estimation.

In Table 1 we have collected some estimates from our experiments. The top line of the table lists the tone-mapping methods. The bottom line gives the running time of every method. The middle lines demonstrate quality values of the used quality measures (they are listed in the first column). The most-right column presents the execution time for mapping transforms and the quality measure calculation respectively. MATLAB programs were used for image processing and running experiments. Some programs have been downloaded from the website of the authors of the mentioned articles and adapted, others were written by myself. They are not optimized and are used only for relative comparison.

Table 1. Tested tone mapping transformations and NR-quality measures

Method, measure	[7]	[6]	[9]	[10]	[10]	[11]	[8loc]	[8glob]	
Our blur measure	**0.170**	0.164	0.159	0.119	**0.170**	0.173	0.160	0.164	0.171
Entropy	6.282	6.299	6.331	2.259	6.547	7.127	**7.201**	6.977	**0.015**
ssim	0.481	0.380	0.402	0.001	0.615	0.001	**0.799**	0.746	0.265
Crete blur	**0.332**	0.329	0.304	0.235	0.330	**0.332**	0.309	0.316	0.218
JNBM, Karam	8.837	9.480	8.550	**14.995**	8.673	8.399	8.983	8.397	2.558
CPBD, Karam	**0.364**	0.328	0.297	0.277	0.297	0.269	0.266	0.278	1.872
TMQI	0.245	0.247	0.256	**0.702**	0.261	0.275	0.262	0.332	0.936
Naturalness	0.026	0.018	0.095	0.000	0.090	0.000	0.145	0.488	0.436
	0.125	0.359	0.359	0.156	0.390	0.171	1.653	0.249	Time, sec

Note: the best parameter values and methods are printed by bold.

Best values in Table 1 are highlighted in bold. From Table one can see that the fastest NR quality measure is entropy, followed by is our blur measure measure which is similar to the Crete measure. Both of them are used low-pass filtering of the initial image and calculate the maximal mean difference between vertical or horizontal variations of two image (input and the blurred one).

Shortly we can say that the Drago variant with gamma correction produce a very bright image, but the JNBM measure of Karam indicates that it is the best variant of tone mapping. The measure has the longest running time. So, it is not a good quality measure. Similar we can reject the TMQI measure developed by Zhu and Milanfar.

Then we tested simple exponential function changing gamma parameter and analyzing behavior of the quality functions. The sharpness index by Blanchet and Moisan increases monotonically and gives the biggest value for biggest gamma, but such DDR-LDR transformation produces almost binary image without any gray values. This means that this sharpness index is a bad measure and we rejected it.

μ-law [7] is the fastest variant of HDR-LDR transformation and there NR quality measures indicate that it is the best variant of tone mapping for SAR images. This evaluation correlate with our visual evaluation also, see Fig. 2.

5   Conclusion

The fastest tone mapping transformation is mu-law, followed by is local variant of Ashikhmin transformation (compare Fig. 2 with Fig. 5). Both of them correlate with the human quality estimation of the transformed images.

Acknowledgment. This research was supported by Belarussian Republican Foundation for Fundamental Research, grant № Ф16СРБГ-004.

References

1. Hisanaga, S., Wakimoto, K., Okamura, K.: Tone mapping and blending method to improve SAR image visibility. IAENG Int. J. Comput. Sci. **38**(3), 289–294 (2011)
2. Lambers, M., Nies, H., Kolb, A.: Interactive dynamic range reduction for SAR images. IEEE Geosci. Remote Sensing Lett. **5**(3), 507–511 (2008). doi:10.1109/LGRS.2008.922732
3. Zhou, G., An, W., Yang, J., Zhong, H.: A visualization method for SAR images. In: IEEE International Geoscience and Remote Sensing Symposium, pp. 374–377. IEEE Press (2011). doi:10.1109/IGARSS.2011.6048977
4. Little, J.R.: Gamma correction circuit, Patent US 2697758 A (1954)
5. Florea, C., Vertan, C., Florea, L.: High dynamic range imaging by perceptual logarithmic exposure merging. Int. J. Appl. Math. Comput. Sci. **25**(4), 943–954 (2015). doi:10.1515/amcs-2015-0067
6. Reinhard, E., Stark, M., Shirley, P., Ferwerda, J.: Photographic tone reproduction for digital images. ACM Trans. Graph. **21**(3), 267–276 (2002). doi:10.1145/566654.566575
7. ITU-R BT.709-5: Parameter values for the HDTV standards for production and international programme exchange, June 2015. https://www.itu.int/rec/R-REC-BT.709/en
8. Ashikhmin, M.: A tone mapping algorithm for high contrast images. In: 13th Eurographics Workshop on Rendering, pp. 145–156. Eurographics Association (2002)
9. Reinhard, E., Devlin, K.: Dynamic range reduction inspired by photoreceptor physiology. IEEE Trans. Visual. Comput. Graph. **11**(1), 13–24 (2005). doi:10.1109/TVCG.2005.9
10. Drago, F., et al.: Adaptive logarithmic mapping for displaying high contrast scenes. Comput. Graph. Forum **22**(3), 419–426 (2003). doi:10.1111/1467-8659.00689
11. Schlick, C.: Quantization techniques for visualization of high dynamic range pictures. In: Sakas, G., Shirley, P., Muller, S. (eds.) Photorealistic Rendering Techniques, pp. 7–20. Springer, Heidelberg (1995). doi:10.1007/978-3-642-87825-1_2
12. Sheikh, H.R., Bovik, A.C.: Image information and visual quality. IEEE Trans. Image Process. **15**(2), 430–444 (2006). doi:10.1109/MSP.2008.930649
13. Ferzli, R., Karam, K.: A no-reference objective image sharpness metric based on the notion of just noticeable blur (JNB). IEEE Trans. Image Process. **18**(4), 717–728 (2009). doi: 10.1109/TIP.2008.2011760
14. Shahid, M., et al.: No-reference image and video quality assessment: a classification and review of recent approaches. EURASIP J. Image Video Process. **2014**(1), 1–32 (2014). doi: 10.1186/1687-5281-2014-40
15. Zhu, X., Milanfar, P.: Automatic parameter selection for denoising algorithms using a no-reference measure of image content. IEEE Trans. Image Process. **19**(12), 3116–3132 (2010). doi:10.1109/TIP.2010.2052820
16. Vu, C.T., Phan, T.D., Chandler, D.M.: S3: a spectral and spatial measure of local perceived sharpness in natural images. IEEE Trans. Image Process. **21**(3), 934–945 (2012). doi:10.1109/TIP.2011.2169974
17. Blanchet, G., Moisan, L.: An explicit sharpness index related to global phase coherence. In: IEEE International Conference on Acoustics, Speech, and Signal Processing, pp. 1065–1068. IEEE Press (2012). doi:10.1109/ICASSP.2012.6288070

18. Leclaire, A., Moisan, L.: No-reference image quality assessment and blind deblurring with sharpness metrics exploiting Fourier phase information. J. Math. Imaging Vis. **52**(1), 145–172 (2015). doi:10.1007/s10851-015-0560-5

19. Gabarda, S., Cristybal, G.: Blind image quality assessment through anisotropy. J. Optical Soc. Am. A **24**(12), B42–B51 (2007). doi:10.1364/JOSAA.24.000B42

20. Crete-Roffet, F., Dolmiere, T., Ladret, P., Nicolas, M.: The blur effect: perception and estimation with a new no-reference perceptual blur metric. In: SPIE Electronic Imaging Symposium Conference on Human Vision and Electronic Imaging, p. EI6492-16 (2007)

21. Sample Radar Imagery: http://www.intelligence-airbusds.com/en/23-sample-imagery.php#

A Shallow Convolutional Neural Network for Accurate Handwritten Digits Classification

Vladimir Golovko[1(✉)], Mikhno Egor[1], Aliaksandr Brich[1], and Anatoliy Sachenko[2]

[1] Brest State Technical University, Moskowskaja 267, 224017 Brest, Belarus
gva@bstu.by
[2] Research Institute for Intelligent Computer Systems,
Ternopil National Economic University, 3 Peremoga Square, Ternopil, 46020, Ukraine
as@tneu.edu.ua

Abstract. At present the deep neural network is the hottest topic in the domain of machine learning and can accomplish a deep hierarchical representation of the input data. Due to deep architecture the large convolutional neural networks can reach very small test error rates below 0.4% using the MNIST database. In this work we have shown, that high accuracy can be achieved using reduced shallow convolutional neural network without adding distortions for digits. The main contribution of this paper is to point out how using simplified convolutional neural network is to obtain test error rate 0.71% on the MNIST handwritten digit benchmark. It permits to reduce computational resources in order to model convolutional neural network.

Keywords: Convolutional neural networks · Handwritten digits · Data classification

1 Introduction

An artificial neural network is powerful tool in different domains [1–5]. Over the last decade the machine learning techniques has the leading role in domain of artificial intelligence [1]. This is confirmed by recent qualitative achievements in images, video, speech recognition, natural language processing, big data processing and visualization, etc. [1–18]. These achievements are primarily associated with new paradigm in machine learning, namely deep neural networks and deep learning [2, 6–18]. However in many real world applications the important problem is limited computational resources, which doesn't permit to use deep neural networks. Therefore the further development of shallow architecture is an important task. It should be noted especially that for many real applications, the shallow architecture can show the comparable accuracy in comparison with deep neural networks.

This paper deals with a convolutional neural network for handwritten digits classification. We propose a simplified architecture of convolutional neural networks, which permits to classify handwritten digits with more precision than a conventional convolution neural network LeNet -5. We have shown that by using a simplest convolutional neural network can be obtained the better classification results.

© Springer International Publishing AG 2017
V.V. Krasnoproshin and S.V. Ablameyko (Eds.): PRIP 2016, CCIS 673, pp. 77–85, 2017.
DOI: 10.1007/978-3-319-54220-1_8

The rest of the paper is organized as follows. Section 2 introduces the standard convolutional neural networks. In Sect. 3 we propose a simplified convolutional network. Section 4 demonstrates the results of experiments and finally Sect. 5 gives conclusion.

2   Related Works

Convolutional neural network is a further development of a multilayer perceptron and neocognitron and is widely used for image processing [19–22]. This kind of neural network is invariant to shifts and distortions of the input. Convolutional neural network integrates three approaches, namely local receptive field, shared weights and spatial subsampling [20–22]. Using local receptive areas the neural units of the first convolutional layer can extract primitive features such as edges, corners etc. The general structure of convolutional neural network is shown in the Fig. 1.

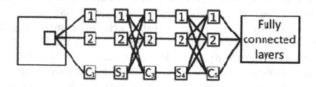

Fig. 1. General structure of convolutional neural network

A convolutional layer consists of set of a feature maps and the neural units of each map contain the same set of weights and thresholds. As a result each neuron in a feature map performs the same operations on different parts of image. The sliding windows technique is used for image scanning. Therefore if size of window is pxp (receptive field) then each unit in a convolutional layer is connected with p2 units of the corresponding receptive field. Each receptive field in input space is mapped into special neuron in each feature map. Then if the stride of sliding window is one that the numbers of neurons in each feature map is given by

$$D(C_1) = (n - p + 1)(n - p + 1) \tag{1}$$

where nxn is size of image. If the stride of sliding window is S that the numbers of neurons in each feature map is defined by the following way:

$$D(C_1) = (\frac{n - p}{s} + 1)(\frac{n - p}{s} + 1) \tag{2}$$

Accordingly, the common number of synaptic weights in convolutional layer is defined by

$$V(C_1) = M(p^2 + 1) \tag{3}$$

where M – the number of feature maps in convolutional layer. Let's represent the pixels of the input image in one-dimensional space. Then the ij-th output unit for k-th feature map in convolutional layer is given by

$$y_{ij}^k = F(S_{ij}^k) \qquad (4)$$

$$S_{ij}^k = \sum_c w_{cij}^k x_c - T_{ij}^k \qquad (5)$$

where $c = 1, p^2$, F – the activation function, S_{ij}^k — the weighted sum of the ij-th unit in k-th feature map, w_{ij}^k — the weight from the c-th unit of the input layer to the ij-th unit of the k-th feature map, T_{ij}^k – the threshold of the ij-th unit of the k-th feature map.

As already said the neural units of each feature map contain the same set of weights and thresholds. As a result the multiple features can be extracted at the same location. These features are then combined by the higher layer using pooling in order to reduce the resolution of feature maps [22]. This layer is called subsampling or pooling layer and performs a local averaging or maximization different regions of image. To this end, each map of convolutional layer is divided into non-overlapping areas with size of kxk and each area is mapped into one unit of corresponding map in pooling layer. It should be noted that each map of convolutional layer is connected only with corresponding map in pooling layer. Each unit of pooling layer computes the average or maximum of k^2 neurons in a convolutional layer:

$$z_j = \frac{1}{k \times k} \sum_{j=1}^{k \times k} y_j$$

$$z_j = \max(y_j) \qquad (6)$$

The number the of neurons in each pooling map is given by

$$D(S_2) = \frac{D(C_1)}{k^2} \qquad (7)$$

The number of feature maps in pooling layer will be the same like in convolutional layer and equal M. Thus convolutional neural network represents combination of convolutional and pooling layers, which perform nonlinear hierarchical transformation of input data space. The last block of the convolutional neural network is a multilayer perceptron, SVM or other classifier (Fig. 2).

Lets consider the conventional convolutional neural network (LeNet-5) for handwritten digits classification (Fig. 3) [22]. The input image has size 32×32. The sliding window with size 5×5 scans the image and the segments of images enter to the layer C1 of neural network. Layer C1 is a convolution layer with 6 feature maps and each feature map contains 28×28 neurons. Layer S2 is a subsampling layer with 6 feature maps and a 2×2 kernel for each feature map. As a result each feature map of this layer contains 14×14 units. Layer C3 is a convolution layer with 16 feature maps and a 5×5

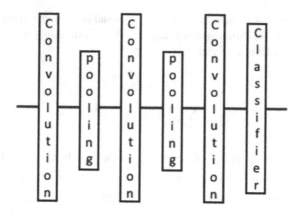

Fig. 2. General representation of convolutional neural network

kernel for each feature map. The number of neural units in each feature map is 10×10. The connections between layers S2 and C3 are not fully connected [22], as is shown in the Table 1.

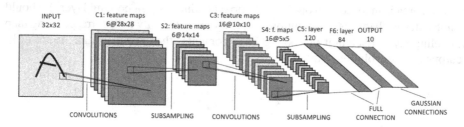

Fig. 3. Architecture of LeNet-5

Table 1. Connections between layers S_2 and C_3

	1	2	3	4	5	6	7	8	9	10	11	12	13	14	15	16
1	X				X	X	X			X	X	X	X		X	X
2	X	X				X	X	X			X	X	X	X		X
3	X	X	X				X	X	X			X		X	X	X
4		X	X	X			X	X	X	X			X		X	X
5			X	X	X			X	X	X	X		X	X		X
6				X	X	X			X	X	X	X		X	X	X

Layer S^4 is a subsampling layer with 16 feature maps and a 2×2 kernel for each feature map. As a result each feature map of this layer contains 5×5 units. Each receptive field with size 5×5 is mapped into 120 neurons of the next layer C^5. Therefore layer C^5 is a convolution layer with 120 neurons. The next layer F^6 and output layer are fully connected layers.

3   The Simplified Convolutional Network

In this section we proposed convolutional neural network which has a simpler architecture compared with LeNet-5. The simplified convolutional neural network for handwritten digit classification is shown in Fig. 4. This network consists of convolutional layer (C^1), pooling layer (S^2), convolutional layer (C^3), pooling layer (S^4) and convolutional layer (C^5). The convolutional layer C^1 has 8 feature maps and each feature map contains 24 × 24 neurons. The pooling layer S2 contains 8 feature maps and 12 × 12 units for each feature map, i.e. k = 2. Layer C^3 is a convolution layer with 16 feature maps and 8 × 8 neurons in each feature map. The layers S^2 and C^3 are fully connected in comparison with conventional network LeNet 5. Layer S4 is a pooling layer with 16 feature maps and 4 × 4 units for each feature map. The last layer C^5 is the output layer contains 10 units and performs classification. As can be seen the main differences are the following: 1) we removed two last layers in LeNet-5 2) the layers S^2 and C^3 are fully connected 3) the sigmoid transfer function is used in all convolutional and output layers. The goal of learning is to minimize the total mean square error (MSE), which characterizes the difference between real and desires outputs of neural network. In order to minimize a MSE we will use gradient descent technique. The mean square error for L samples is defined using outputs of last layer:

$$E_s = \frac{1}{2} \sum_{k=1}^{L} \sum_{j=1}^{m} (y_j^k - e_j^k)^2 \tag{8}$$

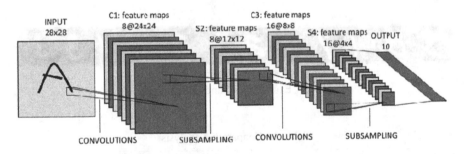

Fig. 4. Architecture of simplified convolutional neural network

where y_j^k and e_j^k – respectively real and desired output of j-th unit for k-th sample. Then using gradient descent approach we can write in case of mini-batch learning, that

$$w_{cij}(t+1) = w_{cij}(t) - \alpha \frac{\partial E(r)}{\partial w_{cij}(t)} \tag{9}$$

where α is learning rate, $E(r)$ is mean square error for r samples (size of minibatch). Since the units of each feature map in convolutional layer contain the same set of weights then the partial derivative $\dfrac{\partial E(r)}{\partial w_{cij}(t)}$ is equal to the sum of partial derivatives for all neurons of the feature map:

$$\frac{\partial E(r)}{\partial w_{cij}(t)} = \sum_{i,j} \frac{\partial E(r)}{\partial w_{cij}(t)} \tag{10}$$

As a result in case of batch learning we can obtain the following delta rule to update synaptic weights:

$$w_{cij}(t+1) = w_{cij}(t) - \alpha(t) \sum_{i,j} \sum_{k} \gamma_{ij}^{k} F'(s_{ij}^{k}) x_{c}^{k} \tag{11}$$

where $c = 1, p^2$, $F'\left(s_{ij}^{k}\right) = \dfrac{\partial y_{ij}^{k}}{\partial S_{ij}^{k}}$ – the derivative of activation function for k-th sample, s_{ij}^{k} – the weighted sum, γ_{ij}^{k} the error of ij-th unit in a feature map for k-th sample, x_{c}^{k} – the c-th input.

4   Experiments

In order to illustrate the performance of proposed technique we present simulation results for handwritten digits classification using MNIST dataset. The MNIST dataset contains 28×28 handwritten digits in gray-scale and has a training set of 60000 samples, and a test set of 10000 samples. You can see some examples of handwritten digits from MNIST data set in the Fig. 5.

Fig. 5. Examples of handwritten digits

Table 2. Conparative analysis

Classifier	Preprocessing	Test error rate (%)
CNN LeNet-1	Subsampling to 16 × 16 pixels	1.7
CNN LeNet-4	None	1.1
CNN LeNet-4 with K-NN instead of last layer	None	1.1
CNN LeNet-4 with local learning instead of last layer	None	1.1
Convolutional net LeNet-5, [no distortions]	None	0.95
Convolutional net LeNet-5, [huge distortions]	None	0.85
Classifier	Preprocessing	Test error rate (%)
Convolutional net LeNet-5, [distortions]	None	0.8
Simplified convolutional net, [no distortions]	None	0.71
Convolutional net Boosted LeNet-4	None	0.7

We used simple backpropagation algorithm for convolutional neural network training without any modifications. The size of mini batch is 50; learning rate is changed from 0.8 to 0.0001. The results of experiments are illustrated in the Table 2. As can be seen we can achieve test error rate 0.71% using simple shallow convolutional neural network. The best result for convolutional network LeNet-5 without distortions is 0.95%. Thus the use of simplified convolutional network with the elementary backpropagation technique permits to obtain the better performance compared conventional architecture. The processing results of each layer of digit 7 are shown in the Table 3.

Table 3. The results of handwritten digits

Layer type	Number and size of map	Processing results (%)
Input	1@28x28	
Convolutional	8@24x24	
Subsampling	8@12x12	
Convolutional	16@8x8	
Subsampling	16@8x8	

5   Conclusion

This paper deals with a convolutional neural network for handwritten digits classification. We propose a simplified architecture of convolutional neural networks, which permits to classify handwritten digits with more precision than a conventional convolution neural network LeNet-5. The main differences from the conventional LeNet-5 are the following: we removed two last layers in LeNet-5; the layers S^2 and C^3 are fully connected; the sigmoid transfer function is used in all convolutional and output layers. We have shown that simple neural network is capable of achieving test error rate 0.71% on the MNIST handwritten digits classification.

References

1. LeCun, Y., Bengio, Y., Hinton, G.E.: Deep learning. Nature **521**, 436–444 (2015)
2. Golovko, V., Bezobrazov, S., Kachurka, P., Vaitsekhovich, L.: Neural network and artificial immune systems for malware and network intrusion detection. In: Koronacki, J., Raś, Z.W., Wierzchoń, S.T., Kacprzyk, J. (eds.) Advances in Machine Learning II. Studies in computational intelligence, vol. 263, pp. 485–513. Springer, Heidelberg (2010)
3. Golovko V., Kachurka P., Vaitsekhovich L.: Neural network ensembles for intrusion detection. In: The 4th IEEE Workshop on Intelligent Data Acquisition and Advanced Computing Systems: Technology and Applications (IDAACS-2007), pp. 578–583, September 2007
4. Dziomin, U.: A multi-agent reinforcement learning approach for the efficient control of mobile robots. In: Dziomin, U., Kabysh, A., Stetter, R., Golovko, V. (eds.) Advanced in Robotics and Collaborative Automation, pp. 123–146. River Publishers, San Francisco (2015)
5. Golovko, V., Artsiomenka, S., Kisten, V., Evstigneev, V.: Towards automatic epileptic seizure detection in EEGS based on neural networks and largest Lyapunov exponent. Int. J. Comput. **14**(1), 36–47 (2015)
6. Hinton, G.E., Osindero, S., Teh, Y.: A fast learning algorithm for deep belief nets. Neural Comput. **18**, 1527–1554 (2006)
7. Hinton, G.: Training products of experts by minimizing contrastive divergence. Neural Comput. **14**, 1771–1800 (2002)
8. Hinton, G., Salakhutdinov, R.: Reducing the dimensionality of data with neural networks. Science **313**(5786), 504–507 (2006)
9. Hinton, G.E.: A practical guide to training restricted Boltzmann machines. (Technical report 2010-000). Machine Learning Group, University of Toronto, Toronto (2010)
10. Bengio, Y.: Learning deep architectures for AI. Found. Trends Mach. Learn. **2**(1), 1–127 (2009)
11. Bengio, Y., Lamblin, P., Popovici, D., Larochelle, H.: Greedy layer-wise training of deep networks. In: Scholkopf, B., Platt, J.C., Hoffman, T. (eds.) Advances in Neural Information Processing Systems, vol. 11, pp. 153–160. MIT Press, Cambridge (2007)
12. Golovko, V., Kroshchanka, A., Rubanau, U., Jankowski, S.: A learning technique for deep belief neural networks. In: Golovko, V., Imada, A. (eds.) Neural Networks and Artificial Intelligence. Communication in Computer and Information Science, vol. 440, pp. 136–146. Springer, Heidelberg (2014)
13. Golovko, V. From multilayer perceptron to deep belief neural networks: training paradigms and application. In: Lectures on Neuroinformatics, pp. 47–84, Moscow (2015)
14. Golovko, V., Kroshchanka, A., Turchenko, V., Jankowski, S., Treadwell, D.: A new technique for restricted Boltzmann machine learning. In: Proceedings of the 8th IEEE International Conference IDAACS-2015, Warsaw, pp. 182–186, 24–26 September 2015
15. Golovko, V., Kroschanka, A.: The nature of unsupervised learning in deep neural networks: a new understanding and novel approach. Opt. Mem. Neural Netw. **25**(3), 127–141 (2016)
16. Golovko, V., Kroschanka, A.: Theoretical notes on unsupervised learning in deep neural networks. In: Proceedings of the 8-th International Joint Conference on Computational Intelligence, NCTA 2016, Porto, Portugal, pp. 91–96, 9–11 November 2016
17. Golovko, V.: Deep neural networks: a theory, application and new trends. In: Proceedings Of The 13th International Conference On Pattern Recognition and Information Processing (PRIP 2016), pp. 33–37. Publishing Center of BSU, Minsk (2016)

18. Golovko, V., Mikhno, E., Brich, A.: A simple shallow convolutional neural network for accurate handwritten digit classification. In: Proceedings of the 13th International Conference on Pattern Recognition and Information Processing (PRIP 2016), pp. 209–212, Publishing Center of BSU, Minsk (2016)
19. Fukushima, K.: Neocognitron: A self-organizing neural network model for a mechanism of pattern recognition unaffected by shift in position. Biol. Cybern. **36**, 193–202 (1980)
20. LeCun, Y., Boser, B., Denker, J., Henderson, R., Howard, R., Hubbard, W., Jackel, L.: Backpropagation applied to handwritten zip code recognition. Neural Comput. **1**(4), 541–551 (1989)
21. LeCun, Y., Haffner, P., Bottou, L., Bengio, Y.: Object recognition with gradient-based learning. In: Forsyth, D.A., Mundy, J.L., di Gesú, V., Cipolla, R. (eds.). LNCS, vol. 1681, pp. 319–345. Springer, Heidelberg (1999). doi:10.1007/3-540-46805-6_19
22. LeCun, Y., Bottou, L., Bengio, Y., Haffner, P.: Gradient-based learning applied to document recognition. Proc. IEEE **86**(11), 2278–2324 (1998)

Highlighting Tumor Borders Using Generalized Gradient

Vassili Kovalev[1](✉), Eduard Snezhko[1], Siarhei Kharuzhyk[2], and Vitali Liauchuk[1]

[1] Biomedical Image Analysis Group and Mathematical Cybernetics Department,
United Institute of Informatics Problems, Surganova Street, 6, 220012 Minsk, Belarus
`vassili.kovalev@gmail.com`
[2] N.N. Alexandrov National Cancer Center of Belarus, 223043 Minsk, Lesnoy District, Belarus

Abstract. This paper presents a generalized approach for computing image gradient. It is predominantly aimed at detecting unclear and in certain circumstances even completely invisible borders in large 2D and 3D texture images. The method exploits the conventional approach of sliding window. Once two pixel/voxel sets are sub-sampled from orthogonal window halves, they are compared by a suitable technique (e.g., statistical t-test, SVM classier, comparison of parameters of two distributions) and the resultant measure of difference (e.g., t-value, the classification accuracy, skewness difference of two distributions etc.) is treated as the gradient magnitude. The bootstrap procedure is employed for increasing the accuracy of difference assessment of two pixel/voxel sets.

Keywords: Lung tumor · CT · Edge detection · Generalized gradient

1 Introduction

In many occasions there is a strong need for detection of borders of objects which look like patterns of random textures. Such borders could be hardly detected by human visual system when textures differ by their high-order statistics only [1, 2]. Recently, some advanced methods of detecting hardly visible borders between the random image textures have been suggested [1, 3]. Moreover, it was experimentally proven that these methods capitalizing on so-called "generalized gradient" are able to highlight the border which is completely invisible for human eye [4, 5].

This paper addresses the problem of detecting borders of malignant tumors in native lung CT images under conditions of presence of atelectasis. The atelectasis term denotes the collapse of all or part of a lung due to bronchial plugging or the chest cavity being opened to atmospheric pressure. This can happen when the vacuum between the lung and chest wall is broken, allowing the lung to collapse within the chest (e.g., pneumothorax), when the lung is compressed by masses in the chest, or when an airway is blocked, leading to slow absorption of the distal air into the blood without replenishment. In this work we were dealing with the bronchial compression caused by lung cancer tumors, the most common cause of the atelectasis.

Computed tomography (CT) is the primary modality for imaging lung cancer patients. However, the problem is that on CT scans the lung regions with the atelectasis and malignant tumors have quite similar attenuation values. Therefore the visual discrimination and

© Springer International Publishing AG 2017
V.V. Krasnoproshin and S.V. Ablameyko (Eds.): PRIP 2016, CCIS 673, pp. 86–96, 2017.
DOI: 10.1007/978-3-319-54220-1_9

separation of the atelectasis and tumor is hardly possible. Yet, accurate tumor segmentation is strongly necessary by the following two reasons. First, the correct tumor localization, segmentation, and precise measurement of tumor diameter play a crucial role in therapy planning and choosing suitable surgery technique. Second, if the radiation therapy is prescribed, an exact separation of tumor border is required for precise targeting and delivery of the ionizing radiation dose accurately to the tumor but not to the surrounding tissues.

Thus, the purpose of this particular paper is to present results of an experimental study of the ability of the generalized gradient method to highlight hardly visible borders of objects. The study was conducted using three different groups of images. They were comprised by 3D synthetic images and specially-designed physical gelatin phantom made by authors and scanned using Siemens Somatom Definition AS scanner. Finally, the utility of the method was examined on the problem of borders detection between malignant lung tumors and the atelectasis regions based on 3D CT images of 40 lung cancer patients.

The first version of the generalized gradient method was introduced in [2] as so-called classification gradient and slightly improved afterwards. The classification gradient method makes use conventional technique of calculating image gradient at each pixel position by means of comparing pixel/voxel values taken from orthogonal halves of appropriately sized sliding window. However, apart from the traditional approaches where the gradient magnitude is computed simply as the intensity difference (estimated by convolution with one or other matrix of weights), the generalized gradient method treats the voxel values taken from window halves as two samples which need to be compared in a suitable way. Once it is done, the value of the corresponding dissimilarity measure is treated as a "gradient" value at the current sliding window position for a given orientation X, Y or Z.

One may prefer to employ a sophisticated technique of comparing two samples of voxels such as the voxel classification procedure performed with the help of an appropriate classifier [3]. In these circumstances the resultant classification accuracy is treated as the local image gradient magnitude which is varied in the range of 0–100%. Along with recent classifiers, the sets of voxels may, for example, be compared in a statistical manner using conventional t-test. This case the resultant t-value is treated as a measure of dissimilarity that is as the signed local "gradient" value.

It should be noted that despite the fact that t-test also compares mean values of two voxel samples, it proceeds in a more correct way taking into account the variances of two distributions. In addition, the t-test has an inherit threshold of significance at $|t| = 1.96$; $p < 0.05$ what is often very problematic to set up in conventional intensity convolutions.

2   Materials

In this study we used three kinds of images containing regions with weak borders which are difficult to detect by human visual system: synthetic 3D images, CT image of the physical gelatin phantom and CT images of chest of 40 patients. Image regions did not form coherent spatial pattern, but rather looked like random textures with difference being the probability density functions of values inside them.

2.1 Synthetic Images

For this experiment, we created a synthetic 3D image with size ($512 \times 512 \times 50$) voxels. Inside this volume a parallelepiped was placed with distances along the corresponding volume margins equal to 128, 128 and 12 voxels. The grey values of the voxels of the inner and outer regions were drawn from two Pearson distributions with different parameters, having the same mean value of $\mu = 200$ and standard deviation $\sigma = 20$, but different skewness values. The inner part was filled with values to have the skewness ω_{in} to be as close as possible to 1 taken throughout all the image slices, and voxels from the outer part were filled with values to have the global skewness $\omega_{out} = 2$.

It should be noted that due to the probabilistic technique of values generation the exact equality of their mean, standard deviation and skewness to the expected ones is hardly possible.

2.2 Physical Gelatin Phantom

The purpose of creating physical phantom was to obtain CT image of some real object, consisting of several adjacent parts with low relative contrast (layers). The phantom was supposed to simulate the commonly encountered problem when objects present on radiological images have barely visible boundaries.

To create such a phantom, we used a cylindrical container filled with several horizontal layers of gelatin. Different levels of CT brightness of each layer were obtained by means of dissolving certain pre-calculated amount of radiocontrast agent Omnipaque in liquid gelatin before its solidification. To control the amounts of radiocontrast agent some provisional measurements of Omnipaque solutions' CT-brightness have been made (see Fig. 1(a) and (b)).

To the amounts of dissolved Omnipaque solution were chosen to increase pure gelatin (reference) CT-brightness by 4, 8, 16 and 32 Hounsfield unit (HU) for different layers relative to the brightness of the reference layer. The reference layer was located at the most bottom of the container. The brightest layer was placed next, then the others (see Fig. 1(c)).

Besides, an additional layer of water with Omnipaque solution introduced was poured to the most top. Thus, one more low-contrast border was made between the upper gelatin layer and the liquid layer.

2.3 Malignant Lung Tumors

In this study, we used 40 CT images of thorax of patients with lung cancer and the atelectasis of a portion of the lung as diagnosed by a qualified radiologist and confirmed histologically. Thirty-three of them were males and remaining seven were females. The age of patients ranged from 41 to 80 years with the mean value of 61.7 years and standard deviation of 8.7 years. CT scanning was performed on a multi-slice Volume Zoom Siemens scanner with the standard clinical kV and mA settings during the one-breath hold. The voxel size of 9 tomograms was in the range of 0.65–0.74 mm in the axial image plane with the slice thickness equal to the inter-slice distance of 1.5 mm. The voxel size of 31 remaining

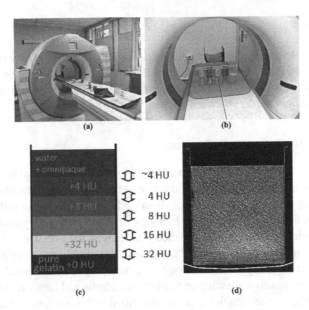

Fig. 1. (a) General view of the installation; (b) cups with different amounts of dissolved Omnipaque solution at the calibration stage; (c) phantom scheme; (d) one slice of the phantom CT image.

tomograms was 0.68 mm in the axial image plane with the slice thickness equal to the inter-slice distance of 5.0 mm. No intravenous contrast agent was administered before the collection of scan data what is a significant detail of present study. Typical examples of original CT image slices are shown in Fig. 2.

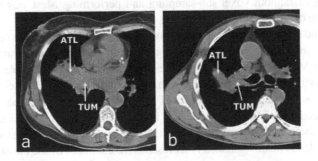

Fig. 2. Example slices of typical lung CT images of two patients with atelectasis (ATL) and malignant tumor (TUM). Patient 1 (left image) suffering from the cancer of middle bronchus with atelectasis of the right middle lobe of the lung. Patient 2 (right image) with the cancer of right upper bronchus and atelectasis of the back segment of the upper lung lobe.

3   Methods

The present study was performed in two main stages. The first, exploratory stage was dedicated to experimental assessment of intensity differences between the regions of malignant tumors and atelectasis. In the second stage we examined the abilities of generalized gradient techniques to highlight borders between the two.

3.1   Exploring the Intensity Differences

The approach followed in this stage was to sub-sample image voxels from two types of lung regions at random and to evaluate the significance of the intensity differences as a function of the sample size (i.e., the number of voxels in each voxel subset). In order to ease the interpretability of the results, the sample sizes were selected so that they correspond to the number of voxels in square-shaped image slice patches with the side size of 3, 4,…, 10, 15, 20, and 30 voxels that is 9, 16,…, 100, 225, 400, and 900 sample voxels respectively. This does not mean that the analysis methodology we developing is 2D-oriented, though. In all the occasions image voxels were sampled from the atelectasis and tumor regions at random. All statistical and pattern recognition analyses described in this work were performed using R, a language and environment for statistical computing which is available for free.

The atelectasis and tumor classes were compared by various ways to eliminate possible bias of one singe method. First, the significance of intensity differences between the two classes was assessed statistically using a two-tailed unpaired t-test with the significance level of t-statistics set to $p < 0.05$. The resultant t-values, which depend on the degree of freedom (sample size) were converted into z-scores to enable direct comparison of statistical significance obtained in different experiments as well as to calculate the mean significance scores over all 40 patients correctly. For each patient and each sample size the procedure consisting of random voxel sub-sampling and performing t-test was replicated 100 times in order to obtain reliable results.

At the second step, the atelectasis and tumor voxel samples (i.e., vectors of voxels sorted in descending order) were clustered using commonly known Hierarchical Clustering, Support Vector Machines, and Random Forests methods. For each sample size and each patient the classifiers were trained on a training sets consisting of 10 atelectasis and 10 tumor samples and tested on the datasets of the same size. Training and test sets were sampled independently. There was no voxels included in both training and test sets simultaneously. The three classifiers were run on exactly the same data. Each test was replicated 100 times in order to obtain statistically representative estimates of the classification accuracy.

The classification accuracy was corrected for agreement by chance using the *classAgreement* function provided with R. For two classes this particularly means that the minimal accuracy value is 0 but not 50%. The corrected classification accuracy was used as a measure of the dissimilarity of two lung regions as well as the basic value for estimating possible image segmentation accuracy. The total number of performed classification tests was: 40 patients × 11 sample sizes × 3 methods × 100 replications = 132 000.

3.2 Detecting Tumor Borders Using Generalized Gradient

The above informal definition of the generalized gradient gives the essence of the method used in present study. The exact computational procedure is a bit more complicated. A list of key details which needs to be considered for better understanding and correct implementation of the method is given below.

Despite the method may be used for computing generalized gradient maps of 2D images, it is better suited for 3D because it is supposed to deal with relatively larger samples of voxels taken from sliding window halves.

It is clear that with no respect to the nature and underlying mechanism of the procedure used for comparing two voxel sets taken from adjacent window halves, it is highly desirable to have the resultant dissimilarity estimate as precise as possible. In order to achieve this, a bootstrap multi-step meta-procedure can be employed (see, for example, a good tutorial [6] written for non-statisticians). In practice it particularly means that at each computational step not the whole amount but a fraction of voxels should be sub-sampled in a random manner from window halves for executing chosen comparison procedure such as t-test. And this step should be repeated about 100 times.

The final dissimilarity measure is computed as a mean value of corresponding particular dissimilarity values that is as the mean t-value computed over the all 100 particular trials in case the t-test procedure is employed. The same holds true in case the final clustering accuracy value is calculated based on particular classification steps, etc. The natural payment for the increased accuracy of assessing the difference by means of bootstrap is the growth of computational expenses for about two orders. For instance, in case of 3D images the total number of elementary t-tests which need to be performed resides around 300 with about 100 tests accomplished for computing gradient components GX, GY and GZ along each of three orthogonal image axes X, Y and Z.

Fig. 3. Configuration of the gap sliding window.

Once the generalized gradient components GX, GY and GZ are computed using the procedure of voxel set comparison, the gradient magnitude $G^{x,y,z}$ at a particular 3D voxel position (x, y, z) is calculated as the Euclidean norm of the vector. In general, the sliding window may have not three orthogonal orientations of voxel sampling like traditional axes X, Y and Z but some alternative configurations too. In this study we also utilized a bit more sophisticated configuration of sliding window depicted in Fig. 3. It supposes to use six directions equally-spaced in 3D. Sampling in each direction is performed using corresponding spherical sub-windows with radius R. Moreover, the sub-windows are moved apart from the central voxel at the distance d. This was done to address the problem of smooth and wide object borders. Finally, the resulting generalized gradient value at a particular 3D voxel position (x, y, z) is calculated from the particular values in each direction $G_i, i \in \{1, \ldots, 6\}$ as $G^{x,y,z} = (\sum_{i=1}^{6} G_i^2)^{1/2}$.

4   Results

4.1   The Intensity Differences Discovered

Results of statistical assessment of the significance of intensity differences between the atelectasis and tumor regions of lung CT scans of 40 patients are reported in Fig. 4. As it can be seen from the figure, the fraction of significance different voxel samples and the mean significance scores varied considerably depending on the patient. For instance, for one patient the percentage of significance different samples exceeds notable 60% already on 9 voxels and achieves 100% with the sample size as little as 36 voxels (see the left panel of Fig. 4) while in other it starts close to zero with 9 voxels and finishes at about 10% only. Similarly, for some patients the mean z-score achieves the significance threshold $z > 1.96$ which is equivalent to $p < 0.05$ on the sample size of 9–25 voxels (see the right panel of Fig. 4) while for others these values remain insignificantly low even on reasonably large samples consisting of 400–900 voxels.

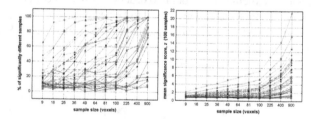

Fig. 4. Significance of the intensity differences of lung atelectasis and tumor voxel samples for 40 patients (curves) as a function of the voxel sample size. Left panel: percentage of voxel samples for which the intensity difference is statistically significant at $p < 0.05$. Right panel: the mean value of significance score z. In both occasions image voxels were sampled from atelectasis and tumor regions at random and each measurement is replicated 100 times.

On the contrary, the voxel sample classification results demonstrate much more consistent behavior (see Fig. 5). As it can be revealed from the figure, a very useful property of the classification approach for separating the atelectasis and tumor regions is that the results are converged to 90–100% of the classification accuracy for relatively large samples in each patient.

As for the comparative efficiency of the three classification methods, it is easy to see from Fig. 5 that the Hierarchical Clustering algorithm outperforms both SVM and Random Forests for each voxel sample size. Moreover, in case of Hierarchical Clustering, the classification accuracy corrected for the agreement by chance starts from the

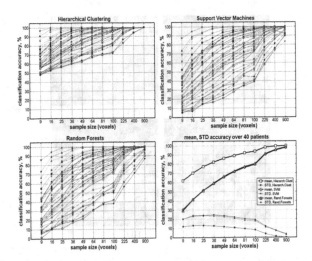

Fig. 5. Dependence of the classification accuracy on sample size of lung atelectasis and tumor voxels for 40 patients (curves) when using Hierarchical Clustering (top left plot), Support Vector Machines (top right plot), and Random Forests (bottom left plot) clustering methods. Each test was replicated 100 times for the reliability of results. The mean and standard deviation accuracy computed over 40 patients is given on the bottom right panel.

value above 50% almost for each patient and achieves 90% on the sample size of 225 voxels for all 40 patients except for 2 outliers. The mean and standard deviation values of the classification accuracy computed over 40 patients (see the bottom right plot of Fig. 5) make the superiority of Hierarchical Clustering method evident and renders other two as almost identical in the voxel sample classification task. Considering that the one possible segmentation technique could be based on a direct voxel sample classification using sliding window of suitable size, the mean accuracy threshold should be set to a reasonably high value, say 95%. If so, the minimal sample size should be set to approximately 100–200 voxels. This corresponds to the window size of about 12×12 voxels (i.e., the half window size is 4.1 mm) for 2D and less than $6 \times 6 \times 6$ voxels (2.0 mm) for 3D case.

4.2 Detected Tumor Borders

The results of application of generalized gradient to synthetic images are depicted on Fig. 6. This experiment shows the capability of the generalized gradient (GG) maps calculated with different presets to detect weak borders, and the results are as they were expected. Figure 6(c) and (d) show the clear border between inner and outer regions. We used the SVM classification accuracy as the difference measure improved by the bootstrap procedure and the gap sliding window. No a priory information about border orientation, width, smoothness or values distribution was used.

Figure 6(b) depicts the GG map calculated over gelatin phantom using conventional t-test to estimate the dissimilarity measure between values sampled from the gap window halves. Though this map calculation is much faster than of the previous ones, in this

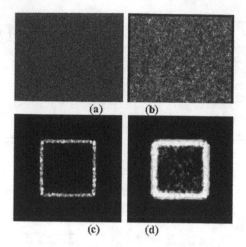

Fig. 6. (a) Original synthetic image; (b) GG map using t-test, $R = 4$, $d = 2$; (c) GG map using SVM, gap window's $R = 3$, $d = 1$; (d) GG map using SVM, $R = 4$, $d = 2$.

particular case it gives no positive outcome, because t-test does not react on the difference of skewness and higher orders moments. However, further we will show that it also provides useful results retaining the same relative advance in speed when used for processing of real images.

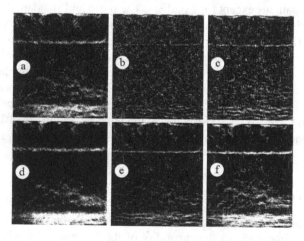

Fig. 7. Gelatin phantom: (a), (d) – GG maps calculated using gap window with $R = 4$, $d = 2$ and $R = 5$, $d = 3$ respectively, t-test of voxel samples used for dissimilarity measure estimation; (b), (e) – GG maps calculated using spherical window with $r = 5$ and $r = 8$ respectively, t-test of voxel samples used for dissimilarity measure estimation; (c), (f) – GG maps calculated using spherical window, dissimilarity measure is the difference of mean values sampled from window halves.

The resultant GG maps of the image in Fig. 1(d) are depicted in Fig. 7. Left column contains maps calculated using t-test to estimate dissimilarity measure and gap sliding window, middle column – also *t*-test and spherical sliding window, right column – spherical sliding window and dissimilarity measure is the difference of mean values sampled from window halves. Sizes of all sliding windows along first and second rows were chosen to have almost the same number of voxels. Unlike the previous synthetic images, the gelatin phantom layers have definite differences of mean HU values, this is why it is fairly easy problem to detect weak borders using different presets of the method. Nevertheless, this figure may help to choose the preferable method's parameters depending on the desired result. The middle layers of gelatin seem to be interdiffused and there were no detectable borders.

Quantitative assessment of the utility of generalized gradient maps in highlighting lung tumor borders was performed separately for the first subgroup of 31 native CT images with the slice thickness of 5.0 mm and remaining 9 images of the second subgroup with the slice thickness of about 1.5 mm. Typical examples of original CT image ROIs and corresponding gradient map regions are presented in Fig. 8.

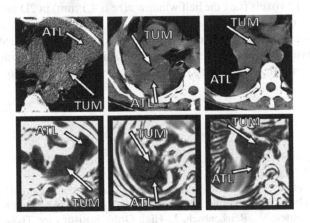

Fig. 8. Example ROIs of the original CT images of lungs (left column) and corresponding generalized gradient maps (right column). The first row represent case where the gradient map is definitely useful for detecting tumor border whereas the second and the third rows illustrate cases where the utility of maps is unclear and useless respectively.

As a result of the experiment, on the first subgroup of patients it was revealed that the generalized gradient maps were definitely useful for detecting tumor border in 17 patients (54.8%) whereas in 9 other cases (29.0%) they did not provide any help for solving the problem of separation the malignant tumor from adjacent atelectasis. The efficacy of maps in the rest 5 cases (16.1%) was found to be unclear. The results of the similar examination of CT scans with reasonably thin slices of about 1.5 mm suggest that it appears to be unlikely the slice thickness is an important parameter for the method. In particular, the distribution of cases between the "yes", "no", and "unclear" categories was 5 (55.6%), 3 (33.3%), and 1 (11.1%) respectively. This is well comparable with corresponding results obtained for the first subgroup.

5   Conclusions

In this work we have documented results of statistical assessment of CT image intensity differences between the lung atelectasis and malignant tumors. The significance scores and classification accuracy results reported here are based on the advanced statistical and pattern recognition methods. Our results suggest that it is unlikely that the use of statistical significance scores for separating lung atelectasis and tumor regions would produce good quality discrimination for all patients. However, the recent clustering algorithms demonstrate some encouraging classification accuracy on the CT intensity samples consisting of few hundred voxels. The Hierarchical Clustering method is found to be better suited for CT voxels classification task comparing to SVM and Random Forests classifiers. This is in agreement with other studies where classes overlap in feature space substantially. The voxel sample classification accuracy potentially allows to reliably discriminate atelectasis and tumor regions using relatively small sliding window of 12×12 voxels (i.e., the half window size is 4.1 mm) in 2D and no more than $6 \times 6 \times 6$ voxels (2.0 mm) in 3D case.

Also we have introduced the basic concept of so-called generalized gradient and demonstrated its abilities and key details on synthetic images, 3D CT images of physical phantom as well as CT scans of lung of 40 patients with clinically confirmed diagnosis of lung cancer.

References

1. Petrou, M., Kovalev, V., Reichenbach, J.: Three-dimensional nonlinear invisible boundary detection. IEEE Trans. Image Proc. **15**(10), 3020–3032 (2006)
2. Kovalev, V., Petrou, M.: The classification gradient. In: Proceedings of 18th International Conference on Pattern Recognition, ICPR-2006, vol. 3, pp. 830–833, Hong Kong (2006)
3. Petrou, M., Kovalev, V., Reichenbach, J.: High Order Statistics for Tissue Segmentation. Handbook of Medical Image Processing and Analysis, 2nd edn, pp. 245–257. Academic Press, San Diego (2009)
4. Wedding, M., Gylys, B.: Medical Terminology Systems: A Body Systems Approach, 5th edn, p. 559. FA Davis Company, Philadelphia (2004)
5. Mira, J., Fullerton, G., Ezekiel, J., Potter, J.: Evaluation of computed tomography numbers for treatment planning of lung cancer. Int. J. Radiat. Oncol. Biol. Phys. **8**(9), 1625–1628 (1982)
6. Wehrens, R., Putter, H., Buydens, L.: The bootstrap: a tutorial. Chemometr. Intell. Lab. Syst. **54**, 35–52 (2000)

Frame Manipulation Techniques in Object-Based Rendering

Victor Krasnoproshin and Dzmitry Mazouka[✉]

Belarusian State University, Minsk, Belarus
krasnoproshin@bsu.by, mazovka@bk.ru

Abstract. This paper is an analysis of rendering methodology based on high-level abstractions. We make a step further from the standard graphics pipeline architecture and include logical objects as a new primitive type into rendering process. The aim of the proposed method is to simplify rendering process definition by reducing the amount of code required to express programming intents. In particular we focus on problems related to frame manipulations and demonstrate the advantages of high-level abstractions over pure pipeline implementation.

Keywords: Graphics pipeline · Rendering · Object-based rendering

1 Introduction

Computer graphics is a very rich technological area, which sees constant improvement on techniques and tools, as well as increasing number of applications. Graphics hardware essentially stays on the cutting edge of parallel computing technologies and rapidly expands its use even to areas unrelated to visualization. However the situation with graphics development complexity does not improve, if we consider standard graphics API, such as DirectX and OpenGL.

The level of abstraction where API cuts off stays very low, for better or worse. On the one hand, the interface remains maximally flexible so it can be applied very broadly, on the other hand, it can be considered too low for any meaningful direct application. This particularity of graphics API has resulted in creation of a vast variety of graphics engines [1] – specialized high-level graphics frameworks providing more convenient visualization abstractions, ready for immediate use.

Graphics engines are great tools for graphics development, however they trade pipeline's flexibility for convenience, and implement often a rigid model of visualization process that can fall short in terms of performance or reasonable applicability under particular circumstances. The most technically advanced engines also remain commercial secrets of corresponding companies. This demonstrates that graphics pipeline and graphics API alone are not enough for high performance and quality visualization, and the accumulated experience could be used to improve the underlying architecture.

© Springer International Publishing AG 2017
V.V. Krasnoproshin and S.V. Ablameyko (Eds.): PRIP 2016, CCIS 673, pp. 97–105, 2017.
DOI: 10.1007/978-3-319-54220-1_10

1.1 Related Works

Improvement of graphics engines architecture has always been in the focus of computer graphics community. Developers from nVidia [2] and Dice [3] recognize the necessity of changing the common scene graph model into something oriented more to the processes of data transformation. This idea primarily comes from the perspective of parallel rendering, which is essentially a cornerstone of graphics hardware, but not so much of a software.

A promising methodology of efficient data-oriented rendering is described by Haaser et al. in "An Incremental Rendering VM" [4]. They show exactly how higher level of abstraction enables the opportunity for automated rendering process optimization.

1.2 Previous Works

In our previous works we studied a theoretical model of visualization process using algebraic abstractions to build a comprehensive set of objects and operations that participate in rendering [5, 6]. In [7] we explored how those constructs could be implemented practically using familiar concepts of high-level shader language (HLSL).

2 Object-Based Rendering

Standard graphics pipeline has a highly robust architecture, which enables vast variety of applications. However, being as extensive as it is, makes the pipeline difficult to use directly for visualization problems. For example, displaying a single rotating 3D object would not be easily called a trivial task when using pure Hardware Abstraction Layer (HAL) functions. The desire for efficiency and flexibility puts a limitation on the programming interface to stay at the low level of abstraction.

Graphics pipeline's operational primitive is a triangle. It is an ultimate building block of any displayed geometry. All objects due for visualization are required to be represented with data arrays of triangles or instructions on how to build those triangles. This data then gets transformed through a number of pipeline stages and then finally visualized on a screen. This fact again makes the pipeline very flexible yet difficult to use in the same sense as assembly programming language is difficult compared to any high level languages.

The problem of graphics pipeline abstraction is commonly addressed by graphics engines. An engine provides a specialized framework that defines data structures and processes that are applied to those structures. The framework typically uses better abstractions and makes applications development much easier. However, graphics engines though reasonably flexible and optimized, tend to conceal underlying HAL functionality and fully build new objects and processes on top of the pipeline. This characteristic of graphics engines is not usually considered a drawback, since the standard abstractions cover very broad number of applications. However, we argue that having specialized engines for visualization problems is not enabling further intensive advancements of graphics pipeline. More studies are required in the direction of bringing higher level abstractions to graphics hardware thus making it easier to use, yet preserving existing flexibility.

In our previous works we defined a notion of visualization algebra. This construct is intended to extend the existing pipeline with a new set of objects and operations. Using a

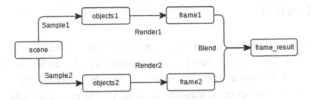

Fig. 1. Rendering process in graph form

generalized representation of objects, we define three basic operations: Sample, Render and Blend. These operations help to build visualization process employing decomposition approach:

1. Sample – selects a subset of objects from a bigger set using a predicate function.
2. Render – a subprocess that performs visualization of a selected set of objects.
3. Blend – composition operation, that merges the outputs of Render or other Blend operations together to form a single frame.

The operations work with two types of algebraic entities: object sets and frames. Object sets originate from the scene – an initial set of objects. Sample operations subdivide the scene and other sets into subsets. Render transforms an object set into a frame. And frames can then be combined using Blend operations. The result of a visualization process is a single frame. It needs to be noted, that operations and corresponding algebraic entities play a descriptive role, defining the intention rather than explicit procedures. That is, visualization process definition can be translated into pipeline instructions sequence in multiple different ways. This allows for automatic process optimization in the very same sense as high level programming language code can be optimized while being translated into assembly instructions.

Rendering process will be formally described using a simple programming notation, having capitalized words defining operations and lower-case words for variables representing algebraic entities described above:

```
objects = Sample(objects)
frame = Render(objects)
frame = Blend(frame1, frame2)
```

Operations used in particular rendering process have to be explicitly defined using, for example, object shaders approach described in our previous work [7].

The whole visualization process is described as a function of scene producing a single resulting frame, for example:

```
frame_result = Blend(
    Render1(Sample1(scene)), Render2(Sample2(scene)))
```

This expression essentially builds a directed graph of data transformation, with nodes being data in various formats and edges – transformation procedures (Fig. 1).

Further in the article we will take a closer look at the Blend operation and frame objects.

3 Frame Algebra

Frame manipulation enabled by Blend operation is the best candidate to demonstrate how visualization process development can be reconsidered in different terms. At the moment, rendering pipeline API does not provide instruments that perform operations on frames explicitly. It is possible, however, to achieve the same results indirectly using render targets and screen-space quad rendering.

For example, let's assume there are two objects that need to be visualized in a way so that their overlapping parts are blended using additive function. We can interpret this problem in terms of frames addition (Fig. 2). The first object is rendered into Frame 1, and the second – into Frame 2, the required result will be a sum of the two frames.

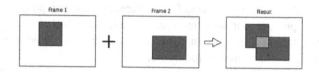

Fig. 2. Overlapping objects rendered with additive blending function

This procedure can be implemented on the graphics pipeline in several ways: Direct approach:

1. Render the first object onto the screen;
2. Set pipeline blending function to Additive;
3. Render the second object onto the screen.

Indirect approach:

1. Set pipeline render target to Texture 1;
2. Render the first object into the render target;
3. Set pipeline render target to Texture 2;
4. Render the second object into the render target;
5. Set pipeline render target to Screen;
6. Set screen space vertex shader;
7. Set pixel shader that adds together two texture channels;
8. Set shader resources to Texture 1 and Texture 2;
9. Render screen quad.

Mixed approach: a combination of any parts from direct and indirect methods.

Depending on particular circumstances any of the provided methods can be applicable during visualization. In simple cases direct approach is the most efficient solution. However more complex visualizations may require rendering into separate textures in order to achieve desired effects. And the example above demonstrates how implementation size can easily grow three times bigger when switching between the methods.

Software development is often a very fluid process and the requirements can drastically change both between and within development projects. Introducing even small differences in initial conditions may result in large changes in the code base, if it is not structured properly. In our case we have a noticeable discrepancy between different implementation methods of the same programming intent – frame addition. If we want to protect against changes in requirements that can impact implementation details, we can use higher level abstractions. In this example the intent would be formulated as short as the following command:

```
frame_result = BlendAdd(
     Render(object1), Render(object2))
```
Now we will compare implementation of a couple of common visual techniques.

Blur. Is a straightforward image filter, that can have various core functions performing the calculation, but the common principle consists just of processing each of the image's pixel and changing it relative to its surrounding. In the terms of frame algebra, the blur function takes one argument and returns a changed value (Fig. 3).

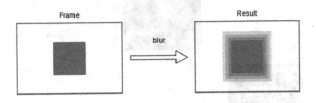

Fig. 3. Blur frame operation.

Pipeline implementation of this operation may look in the following way:

1. Set render target: Texture1
2. { Draw objects }
3. Set screen space vertex shader
4. Set blur pixel shader
5. Set texture input: Texture1
6. Set render target: Screen
7. Draw screen quad

That is, scene objects are rendered first into a separate texture, which then gets rendered onto the screen using an image processing shader that defines blur function. It is easy to notice how pipeline instructions resemble assembly code, in the same way pipeline has a number of dedicated registers and basic operations on them. In this case frame-based implementation will consist just of a single call:

```
frame_result = BlendBlur(Render(objects))
```

Of course, blur blending function needs to be defined with pipeline instructions before use:

1. Set screen space vertex shader
2. Set blur pixel shader
3. Set texture input: argument

4. Draw screen quad

And even though there is not much to be gained in terms of code size, mere calling notation helps to automate the construction of an intermediate texture object and redirect rendering there, before using the texture in image processing.

Bloom. Is a slightly more complicated technique and it usually goes together with High Dynamic Range rendering. The goal of the effect is to emphasize bright areas of the screen by increasing their perceived luminosity. This is done in order to mimic physical processes happening in optical parts of recording devices and human eyes (Fig. 4). The algorithm of the technique consists of two steps: at first the whole scene is displayed on the screen, then bright parts of the scene get rendered into a separate texture that gets blurred and then additively blended with the screen.

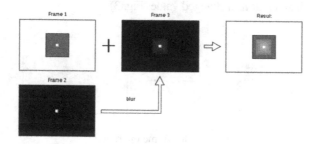

Fig. 4. Bloom effect.

Pipeline implementation of the effect:

1. { Draw objects }
2. Set render target: Texture1
3. { Draw emitting objects }
4. Set screen space vertex shader
5. Set blur pixel shader
6. Set texture input: Texture1
7. Set render target: Screen
8. Set blend function: addition
9. Draw screen quad

It may not be clear, given the pipeline implementation, what exactly is the programming intention here. In the same way as assembly code representing a small algorithm having a few separate steps can seem quite cryptic, and requiring an effort to understand. Comparing this pipeline code with corresponding frame-based notation, it is needless to say that it lacks those deficiencies:

```
frame_result = BlendAdd(Render(objects),
    BlendBlur(Render(SampleEmitters(objects))))
```

Where SampleEmitters is a sample operation that selects objects of Emitter type from the object buffer.

Shadow Maps. This technique solves another common visualization problem: displaying shadows cast by objects from light sources. Basic implementation of this technique requires creation of an auxiliary texture called DepthMap from the light's point of view, containing the distance of lit objects' fragments from the light source. This texture is then used in a separate rendering step when rendering process calculates the contribution of light to screen colours. The scheme of the algorithm is depicted on Fig. 5.

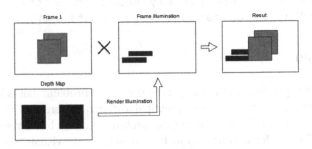

Fig. 5. Shadow Map technique scheme.

Pipeline implementation:

1. Set render target: DepthMap
2. { Draw objects depth map from light perspective }
3. Set render target: FrameIllumination
4. Set texture input: DepthMap
5. { Draw objects illuminated by light }
6. Set render target: Screen
7. { Draw objects }
8. Set screen space vertex shader
9. Set screen pixel shader
10. Set texture input: FrameIllumination
11. Set blend function: multiplication
12. Draw screen quad

The differentiating aspect of this technique is that it uses a frame as an object in another Render procedure. Thus an auxiliary frame needs to be prepared at some moment before the rest of rendering is performed on the pipeline. This complicates instruction sequence even more, as now we can have frame operations mixed with rendering sequences. Obviously, an appropriate high-level representation of this process would look like:

```
depth_map = RenderDepth(objects, light)
illumination = RenderIllumination(
     objects, light, depth_map)
frame_result = BlendMultiply(
     Render(objects), illumination)
```

An addition of a light source in this case would be quite cumbersome on the pipeline, yet using frame algebra it would be as simple as this:

```
depth_map1 = RenderDepth(objects, light1)
illumination1 = RenderIllumination(
    objects, light1, depth_map1)
depth_map2 = RenderDepth(objects, light2)
illumination2 = RenderIllumination(
    objects, light2, depth_map2)
frame_result = BlendMultiply(Render(objects),
    BlendAdd(illumination1, illumination2))
```

4   Discussion

It is fair to say that graphics engines target the very same problem that is in the focus of object-based rendering, however the principal difference here lies in fundamental approach to the problem. Most of the engines architectures are built around a model of scene graph. This model is supposed to fully describe any visualization problem in common terms, thus an optimal solution for the model would also apply to any visualization. However, as convenient as scene graph is, it makes too large a leap in abstraction from standard pipeline API, making engines implementations vary tremendously. A lot of efforts are invested into development of generic graphics engines, and when they are used in application to real problems there is often a risk of hitting their limitations. In this situation it is important to select the right engine for the task, or adapt the task to available engine, or try and extend the engine to accommodate extra requirements. These all are the signs of inflexibility that takes roots at the missed abstraction layer between graphics API and scene graph.

Object-based approach on the other hand does not claim the place of graphics engine, it could rather become a basis for one. The method iterates on pipeline principles and provides a couple of useful abstractions that help to automate visualization processes construction. It formalizes and implements the terms that are already used in graphics development not introducing anything radically new. And more importantly, object-based rendering preserves flexibility of the graphics API and does not stay in the way for trivial problems that do not require higher abstractions.

There are a lot of opportunities for automated optimization in object-based rendering and Haaser et al. [4] propose a promising method called the incremental virtual machine, that compiles sequences of rendered objects into machine instructions. This compilation is performed in the runtime and in the way as to ensure efficient execution on the pipeline.

Object-based rendering in comparison to incremental rendering VM works with different entities. Algebraic constructions and operations aim to manipulate data streams and build the order of processing flow in accordance with programming intent. As soon as data is resolved in form of plain rendered object sequences, incremental rendering can fill the gap between the abstraction and pipeline instructions in the most efficient way.

It is worth noticing that modern graphics APIs are already making some transformations in the direction of higher process description abstractions. The prominent example is Vulkan API [8]. This API introduces multistaged customizable graphics pipeline objects that can be bound to command buffers and then executed. This is very close to object-based rendering process construction approach, however Vulkan API still operates on the level of triangles and surface primitives. It does not support a notion of objects and does not operate with object-based data flows. But it is clear that graphics development as a whole is moving in this direction.

5 Conclusion

This paper provides a detailed look into one of the aspects of object-based rendering methodology. We demonstrate how high-level abstractions can benefit graphics development and where do they fit in the complete scheme of computer visualization. Our goal is to provide developers with appropriate tools that do not compromise on graphics pipeline's flexibility, and make room for automated optimization.

With this work we hope to open a discussion on the future of graphics API evolution.

References

1. Graphics Engines Database, DevMaster (2016). http://devmaster.net/devdb/engines
2. Tavenrath, M., Kubisch, C. (NVIDIA): Advanced scenegraph rendering pipeline. In: GPU Technology Conference, San Jose (2013)
3. Andersson, J., Tartarchuk, N. (DICE): Frostbite rendering architecture and realtime procedural shading texturing techniques. In: Game Developers Conference, San Francisco (2007)
4. Haaser, G., Steinlechner, H., Maierhofer, S., Tobler, R.F.: An incremental rendering VM. In: Proceedings of the 7th Conference on High-Performance Graphics, HPG 2015, Los Angeles, pp. 51–60 (2015)
5. Krasnoproshin, V., Mazouka, D.: Graphics pipeline automation based on visualization algebra. In: 11th International Conference on Pattern Recognition and Information Processing, Minsk (2011)
6. Krasnoproshin, V., Mazouka, D.: Novel approach to dynamic models visualization. J. Comput. Optim. Econ. Finance 4(2–3), 113–124 (2013). New York
7. Krasnoproshin, V., Mazouka, D.: Data-driven method for high level rendering pipeline construction. In: Golovko, V., Imada, A. (eds.) ICNNAI 2014. CCIS, vol. 440, pp. 191–200. Springer, Heidelberg (2014). doi:10.1007/978-3-319-08201-1_18
8. Vulkan, The Khronos Group Inc. (2016). https://www.khronos.org/vulkan/

Pattern Recognition: Supervised Learning on the Basis of Cluster Structures

Rodchenko Vadim[✉]

Yanka Kupala State University of Grodno, Grodno, Republic of Belarus
rovar@grsu.by

Abstract. The original method of recognition of the hidden regularities in data of the training set is described in the paper. The method is based on: the analysis of properties combinations of a space describing objects; the construction of cluster structures; the search for sub-spaces where patterns of classes do not intersect. Application of this method for the solution of problems of pattern recognition with supervised learning is shown.

Keywords: Pattern recognition · Supervised learning · Cluster structure

1 Introduction

The development of recognition systems necessitates solving three main problems. The first problem is connected with representation of basic data and generation of the aprioristic dictionary of features which rather fully describe patterns of classes. The second problem is identification of informative features providing the correct classification of the objects. Finally, the third problem is connected with creation of the qualifier [1].

Features of the aprioristic dictionary possess different degrees of informational content from the point of view of recognition. If, as a result of the solution of the second problem, it is possible to find features or their combinations which provide divisibility of classes, then creation of the qualifier doesn't cause special difficulties [2].

In the article we propose the supervised learning method to create a subspace of features in which classes are divided. This method is based on the analysis of values and the relations between features of the training set in the automatic mode. The possibility of use of this method for the development of pattern recognition systems is shown.

2 Two Approaches to Realization of Supervised Learning

In machine learning we distinguish two types of learning: deductive and inductive. Deductive learning is directed to formalization of knowledge of experts for the purpose of their computer representation in the form of knowledge bases. Inductive learning is based on detection of regularities in empirical data [3, 4].

Further we assume that on the basis of the saved-up empirical data the training set is constructed, i.e. the problem of supervised learning is considered. Analyzing the prior

© Springer International Publishing AG 2017
V.V. Krasnoproshin and S.V. Ablameyko (Eds.): PRIP 2016, CCIS 673, pp. 106–113, 2017.
DOI: 10.1007/978-3-319-54220-1_11

works connected with the supervised learning problem it is possible to allocate two main approaches to its solution.

In the majority of works the problem of learning comes down to creation of the decisive rules providing an extremum for some criterion (for example, for criterion of average risk in a special class of decisive rules). At the same time our attention isn't focused on the problem of formation of feature dictionary for the description of objects. The class of decisive rules is set to parameters in advance, and the problem of learning comes down to finding of values of parameters providing an extremum for the chosen criterion [5]. It is actually supposed that the space of the description of objects is a priori set, and in this space it is necessary to create the dividing surface within the chosen criterion.

Alternative approach to the learning is based on the ideas of construction of feature space in which classes aren't crossed. In this case the procedure of classification of the studied object within the chosen criterion becomes trivial [6].

In this paper we propose the supervised learning method in which on the basis of the analysis of values and the relations of features of training set feature spaces are defined in which patterns of classes aren't crossed.

3   Values and Relations of Training Set Features

Let there is a selection of objects, each of which is the representative of a certain class and is formally described as a vector column $z^T = (z_1, z_2, ..., z_n)$, where n – a quantity of features of the aprioristic dictionary, z_i – a value of i-th feature. Association of all objects of classes is determined by training set which can be formally presented in the form of a matrix of $Z_{n \times m}$, where n – a quantity of signs of the aprioristic dictionary, $m = m_1 + m_2 + ... + m_k$, and m_i – a quantity of objects of i-th class, k – a quantity of classes.

Let's carry out some normalization of the training set $Z_{n \times m}$ (where $m = m_1 + m_2 + ... + m_k$, and m_i - quantity of objects of i- class) and we will get $X_{n \times m}$, where $x_{ij} = (z_{ij} - z_{min})/(z_{max} - z_{min})$. All vectors of i- class will be represented with a separate matrix:

$$X^i_{n \times m_i} = \begin{pmatrix} x^i_{11} & x^i_{12} & \cdots & x^i_{1m_i} \\ x^i_{21} & x^i_{22} & \cdots & x^i_{2m_i} \\ \cdots & \cdots & \cdots & \cdots \\ x^i_{n1} & x^i_{n2} & \cdots & x^i_{nm_i} \end{pmatrix}, \text{ where } i = \overline{1, k}; j = \overline{1, m_i}.$$

Each object of a class in space R^n is represented by a vector with top coordinates $(x_1, x_2, ..., x_n)$, where $x_i \in [0,1]$ – value of i-feature.

It is suggested, on the basis of the analysis of data of the training set, creating such signs subspace in which patterns of classes aren't crossed. The situation when it is possible to create such subspace quickly, is unlikely, but isn't excluded. The example of such a situation is when for one or several features from the aprioristic dictionary it turns out that intervals of changes of values of a feature aren't crossed for all possible couples of classes. However in practice it can occur extremely rarely, and therefore we will pass to consideration of cases when values of feature in the training set submit to more complex distribution laws.

Let there are three classes of objects (i.e. m = 3), each of objects is described by two features (i.e. n = 2), and their distribution in classes is provided in the Fig. 1:

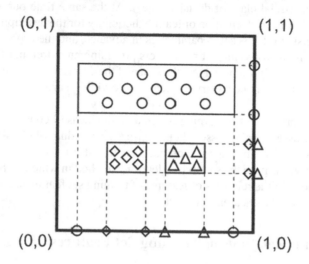

Fig. 1. Example of distribution of objects of three classes

It is obvious that objects of three classes are distinctly divided in two-dimensional feature space, and the problem of classification of a new object looks trivial. If we use only numerical methods and carry out the analysis of values of each feature separately, then: (1) on abscissa axis the classes "rhombuses" and "triangles" are divided among themselves, whereas the class "circles" blocks them; (2) on ordinate axis intervals of change of values of the classes "rhombuses" and "triangles" coincide, and division goes on the class "circles". In this case it isn't simple any more to create an algorithm which on the basis of the analysis of only values of features "will prove" that in this task classes are divided.

From this example it is possible to make some conclusions that the analysis of data of the training set has to consider not only values of features, but also the relations between them.

Let's consider another example. Let in two-dimensional feature space objects of two classes be distributed as follows:

It is obvious that intervals of change of values of features in the classes "circles" and "rhombuses" either practically coincide (on abscissa axis), or on three quarters are crossed (on ordinate axis). And if in the first case (Fig. 1) it is possible to offer option of search of combinations of intervals for division of patterns of three classes, then in the second (Fig. 2) - such approach won't yield positive results. Obviously, geometrical division of classes in this case can be reached only by the account of signs couples "value – relation".

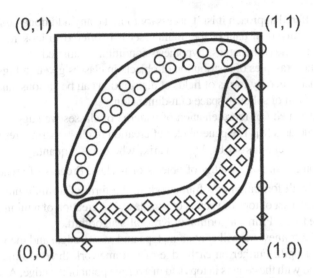

Fig. 2. Example of distribution of objects of two classes

Having increased dimension of space to n = 3, it is easy to imagine the situation when objects of two classes are accurately divided and distributed around cube diagonal tops, and at any projection on the verge of a cube (i.e. in space of dimension (2) or on cube edges (i.e. in space of dimension (1) there is no division of objects of classes. It demonstrates that if there is n of features, then space of the smallest dimension in which classes can be divided has dimension of p, where p ≤ n.

As a result of the carried-out reasonings it is possible to make two important conclusions:

- without account of couples "value – relation" of the training set features it is possible to realize only the "superficial" analysis of data. In this case it is possible to claim that with high probability the available versions of the solution will be missed;
- patterns of classes can be divided in space of dimension p, where p ≤ n, and at the same time there won't be division in spaces of a smaller dimension.

4   Description of the Method of Detection of the Hidden Regularities on the Basis of Cluster Structures

The central point in the solution of the problem of supervised learning is allocated to a compactness hypothesis which says that in space of features compact sets correspond to patterns of classes [7]. It is known that this hypothesis is not always confirmed experimentally. For example, if at the description of objects there are non-informative features, then patterns of one class can be far apart and disseminated among other patterns.

In all the methods in which the problem of learning comes down to creation of the decisive rules providing an extremum of some criteria, as a rule, an additional assumption is used. The assumption is that in space of features the compactness hypothesis is necessarily confirmed.

Within the offered approach it isn't necessary to make any additional assumptions. On the basis of the analysis of data of the training set it is offered to look for such features subspaces in which the compactness hypothesis condition is satisfied.

Analyzing the examples of distribution of objects in classes given in Figs. 1 and 2 it is easy to notice that generally forms of fields of distribution can be various, and areas can be presented in the form of solids in space of n-dimension.

As universal remedy of representation of patterns of classes we suggest using the so-called cluster structures [8]. The general idea of creation of cluster structures is as follows.

Formally each i- class is set by $X^i_{n \times m_i}$ matrix, where n – a quantity of features of the aprioristic dictionary, m_i – a quantity of objects of i- class. Process of creation of cluster structure begins with formation of a framework in the form of a minimum spanning tree which is created at a set of tops of vectors of i-th class. For creation of a minimum spanning tree it is possible to use Prim's algorithm or Kruskal's algorithm.

As a result the framework will contain m_i tops and $m_i - 1$ edge, and the weight of each edge will be its length. Further on each edge of a framework three n-dimensional hyper spheres are placed with the centers in tops and in average point of the edge. At the same time the basic element of cluster structure will have the following appearance (Fig. 3):

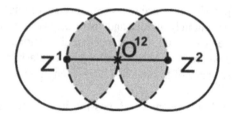

Fig. 3. Basic element of cluster structure

Process of formation of cluster structure comes to the end when all basic elements constructed on the basis of all edges of the earlier created minimum spanning tree are gathered (Fig. 4).

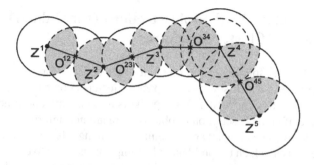

Fig. 4. Example of cluster structure

The cluster structure constructed in such way sets a solid in n-dimensional features space. It, obviously, can be used for formalized representation of a pattern of a class.

Let the training set be represented by matrix $X_{n \times m} = \bigcup_{i=1}^{k} X^i_{n \times m_i}$, where n – a quantity of features of the aprioristic dictionary, $m = m_1 + m_2 + \ldots + m_k$, and m_i – a quantity of objects of i- class, k – a quantity of classes, $X^i_{n \times m_i}$ – a matrix constructed on the basis of all objects of i- class. Let's check the hypothesis of existence of such space of dimension p (where $p \leq n$) in which classes are divided.

The algorithm of check of the hypothesis in this case will be at the same time the algorithm of the "profound" analysis of data for searching the spaces in which classes are divided. Such algorithm can be described in the form of the sequence of the following steps:

Step 1: Let $p = n$. According to the training set $X_{n \times m} = \bigcup_{i=1}^{k} X^i_{n \times m_i}$ we build k of cluster structures. Each i- cluster structure is created on the basis of data of the corresponding matrix $X^i_{n \times m_i}$. If it turns out that all cluster structures don't intercross among themselves, then learning was successful, and created cluster structures are patterns of classes in space of dimension n. Otherwise we suppose $p = 2$ and we pass to the following step of the algorithm.

Step 2: For all possible combinations about p of features from n we look for and we remember combinations of features in case of which cluster structures corresponding to classes don't intercross. In total we analyze C^p_n possible combinations of features and then we pass to the following step of the algorithm.

Step 3: We increase value p on 1 and if $p \leq n - 1$, then we return to the previous step of the algorithm, otherwise we finish it.

Step 2 in the specified algorithm is executed $L = \sum_{i=2}^{n-1} C^i_n$ once. As a result of its realization can be found $0 \leq$ by $L^* \leq L$ of feature subspaces in which patterns of classes aren't crossed.

Every subspace found in such way possesses the following property. Actually combinations of features of a subspace describe the hidden regularity of the kind: *"in feature space created on the basis of the corresponding combination of features, patterns of classes aren't crossed"*.

Model of such a hidden regularity is the single hyper cube in a subspace of features in which the cluster structures are placed which were received as a result of algorithm execution. Important property of the revealed combinations of features is the possibility of their interpretation within the analyzed subject domain.

In conclusion we note that there can be a situation when there is no subspace, in which patterns are divided, i.e. $L^* = 0$. In this case it is necessary to reform the training set. For example, it can be made on the basis of combinations of features of the classes of cluster structures corresponding to patterns providing the minimum crossing. Another option is to create the training set on the basis of a new feature set and apply once again the algorithm described above.

5 Application of the Developed Method in Pattern Recognition

Process of pattern recognition provides consecutive performance of the following stages of processing:

1. preparation of initial data set on purpose of formation of the training set;
2. data preprocessing for elimination of the problems connected with admissions, noise, abnormal values in data, with its redundancy or insufficiency;
3. transformation and normalization of data with the purpose of approaching of data to representation, suitable for the further analysis;
4. supervised learning for identification of the subspace of features providing division of patterns of classes;
5. construction on the basis of the revealed combinations of features of the qualifier and application of the nearest neighbor search for classification of objects.

It is obvious that the result of performance of three initial stages will be the training set which at the fourth stage is subjected to the analysis with use of the supervised learning method described above and is received L^* combinations of features (where $0 \leq L^* \leq L$).

As for each revealed combination of feature patterns of classes in feature space aren't crossed, qualifiers are created on the basis of the nearest neighbor method.

6 Conclusion

In the paper the original method of automatic supervised learning is proposed. This method is based on construction of feature space in which patterns of classes aren't crossed.

For separation of the informative (from the point of view of recognition) features it is proposed researching various combinations of features from the aprioristic dictionary.

The brute-force search algorithm of spaces of partitioned classes which is based on use of cluster structures both at a stage of profound data analysis of training set, and in case of creation of model of representation of the revealed regularities is described.

The proposed algorithm can be parallelized that will allow to reduce significantly time of computation in case of solution of application-oriented tasks. Decomposition of the algorithm on fragments of computation executed in parallel can be realized at the level of a separate combination of features.

The possibility of use of the proposed method to develop pattern recognition systems is shown.

References

1. Tou, J., Gonzalez, R.: Pattern Recognition Principles. Publishing House "Mir", Moscow (1978). (in Russian)

2. Braverman, E.M.: Structural Methods of Handling of Empirical Data. Publishing House "Nauka", Moscow (1983). (in Russian)
3. Flakh, P.: Machine Learning. Science and Art of Creation of Algorithms Which Take Knowledge from Data. Publishing House "DMK Press", Moscow (2015). (in Russian)
4. Dyakonov, A.G.: Analysis of Data, Learning in Precedents, Logical Games, WEKA Systems, RapidMiner and MatLab. Publishing Department of Faculty of VMK Lomonosov Moscow State University, Moscow (2010). (in Russian)
5. Zagoruyko, N.G.: Applied Methods of the Analysis of Data and Knowledge. Publishing Department of Institute of Mathematics, Siberian Branch of the Russian Academy of Science, Novosibirsk (1999). (in Russian)
6. Vasilyev, V.I.: Problem of Learning of Pattern Recognition. Publishing House "Vyscha Shkola", Kiev (1989). (in Russian)
7. Professional Information and Analytical Resource Dedicated Machine Learning, Pattern Recognition and Data Mining. http://www.machinelearning.ru/wiki/index.php?title=Метод_ближайших_соседей. (in Russian)
8. Krasnoproshin, V.V., Rodchenko, V.G.: Cluster structures and their application in data mining. J. Inform. 2(**50**), 71–77 (2016). (in Russian)

Reduction of the Spectral Feature Space Dimension in the Multiclass Problem of ECG-Signals Recognition

L. Manilo and A. Nemirko[✉]

Saint Petersburg Electrotechnical University "LETI", Saint Petersburg, Russia
{lmanilo,apn-bs}@yandex.ru

Abstract. The problem of multiclass recognition of ECG signals using ECG description in the frequency domain is considered. The basis for developing the recognition algorithm is the analysis of Fisher's linear discriminant. Experiments on real signals showed a high efficiency of dangerous arrhythmias recognition.

Keywords: ECG-signal analysis in frequency domain · Fisher linear discriminant · Arrhythmia recognition

1 Introduction

The most important tasks for cardiological monitoring of ECG signals are detection of dangerous arrythmias (ventricular fibrillation and ventricular fluttering) at the time of its sudden occurrence, as well as detection of anomalies, which predict a patient's grave condition. These arrhythmias are: polytopic ventricular extrasystoles, paroxysmal tachycardia, bidirectional ventricular tachycardia (polytopic ventricular tachycardia – torsades de pointes). It causes the problem of creating decision functions for many classes of electrocardiographic signals. The most frequently used is signal description in the frequency domain, which is highly informative for the detection of dangerous arrhythmias [1, 2].

Recognition of the electrocardiographic signals in the frequency domain as a rule is based on the spectral characteristics obtained by calculation of the power spectral density (PSD) function. This description adequately reflects the frequency-response properties of the presented groups of signals, but involves the necessity of determining discriminant functions in a larger dimensional space, and exacerbates creation of the classification algorithms. It may be possible to reduce the feature-space dimension by grouping the spectral coefficients, as well as by mapping the obtained description to a smaller dimensional space using multiple discriminant analysis. The decision functions are created on the basis of analysis of Fisher's linear discriminant J, maximizing of which provides the best choice for the separation of c-classes in the set of signal vectors $(c-1)$ [3]. However, its optimization does not provide reliable detection of signals in every case.

The J criterion for assessing the separation degree of the signals' initial classes can be represented by a scalar value defined by the trace of matrix as follows

$$J = tr(\mathbf{S}_2^{-1}\mathbf{S}_1), \tag{1}$$

© Springer International Publishing AG 2017
V.V. Krasnoproshin and S.V. Ablameyko (Eds.): PRIP 2016, CCIS 673, pp. 114–118, 2017.
DOI: 10.1007/978-3-319-54220-1_12

where S_1 is the scattering matrix between the classes, S_2 is the generalized scattering matrix inside the classes.

For c-classes, in case of transition from the L-dimensional space formed by the spectral features $G = (G_1, G_2, \ldots, G_L)^T$ into the $(c-1)$-dimensional space projections of the objects can be obtained through the following matrix transformation: $Y = W^T \cdot G$, where W is a matrix of dimension $L \times (c-1)$, obtaining of which involves maximization of J. The disadvantage of formula (1) is due to the fact that increasing in the classes' quantity turns criterion J into the indicator of large distances between the groups, which results in poor reflection of mutual locations of closely located classes in the frequency domain with the J criterion.

2 Optimization of the Decision Rules Creation

The procedure of decision rules creation can be optimized by reducing it to a set of pairwise classification tasks and introducing weighting coefficients to increase the impact of closely spaced classes on the J criterion [4]. In this case the generalized formula for the J criterion takes the form:

$$J = \sum_{i=1}^{c-1} \sum_{j=i+1}^{c} n_i n_j a_{i,j} \cdot tr\left[(W^T S_2 W)^{-1} (W^T S_1^{(i,j)} W) \right], \tag{2}$$

where n_i and n_j are rates of occurrence of the objects that form classes ω_i и ω_j.

Weight function $a_{i,j}$ can be linked to the recognition error values of each pair of classes ω_i and ω_j. Publication [4] proposes using weights in the form of a certain error function representation $erf\left(\dfrac{\eta - t}{\sigma}\right)$, where t is the decision rule limit, while η and σ are parameters of the distribution, calculated for specified groups of the objects based on the assumption of the normal distribution law with equal covariance matrices. This approach seems to be effective, due to the possibility of approximating the criterion J to the evaluation of object detection reliability by summing the probabilities of correct solution within pairwise classification. This publication develops the idea of the criterion J approximation to the object classification accuracy in the space of the spectral features that are presented by the standardized values of the PSD.

Determining the elements of matrix W is reduced to the eigenproblem for the following matrix:

$$S_2^{-1} \cdot \sum_{i=1}^{c-1} \sum_{j=i+1}^{c} n_i n_j a_{i,j} \cdot S_1^{(i,j)}$$

and taking columns of the matrix W with dimension $L \times d, (d = c - 1)$ as equal to d eigenvectors that correspond to d maximum eigenvalues λ_i of them.

It is proposed that the task of determining $a_{i,j}$ in (2) as described below be solved.

3   Weight Function Calculation Method

Let us consider two classes of objects ω_i and ω_j with the normal law of distribution and unity matrices of covariation within the two-dimensional space (x_1, x_2). If the distance between the centers of these classes is designated as $\Delta_{i,j} = \|\mathbf{m}_i - \mathbf{m}_j\|$ where $\mathbf{m}_i$ and $\mathbf{m}_j$ are vectors of average values, then projection of these classes on the new direction $\mathbf{V}$ changes the distance between them depending on the angle α between the direction which joins centers of the classes, and the $\mathbf{V}$ vector. This dependence can be represented as $\Delta_{i,j}^{(v)} = \Delta_{i,j} \cdot \cos \alpha$. With identical a priori probabilities of the both class objects occurrences, the probability of the correct recognition is determined as follows:

$$\gamma_{i,j} = \frac{1}{2} + \gamma_{i,j}' = \frac{1}{2} + \frac{1}{2} \cdot erf\left[\frac{\Delta_{i,j}^{(v)}}{2\sqrt{2}}\right],$$

where $erf(.)$ is the error function.

Then, for the case of c-classes with identical distributions, the criterion $J^{(\gamma)}$ that evaluates the average accuracy of recognition can be represented as follows:

$$J^{(\gamma)} = \sum_{i=1}^{c-1} \sum_{j=i+1}^{c} n_i n_j \gamma_{i,j} \tag{3}$$

and the criterion J, which evaluates the range of divergence between the classes can be represented as follows:

$$J = \sum_{i=1}^{c-1} \sum_{j=i+1}^{c} n_i n_j a_{i,j} \cdot tr\left[\left(\mathbf{V}^T \mathbf{S}_1^{(i,j)} \mathbf{V}\right)\right] \tag{4}$$

Comparison of (3) and (4) gives weight values $a_{i,j}$ as follows:

$$a_{i,j} = \gamma_{i,j}'/tr\left(\mathbf{V}^T \mathbf{S}_1^{(i,j)} \mathbf{V}\right)$$

for the case of the best mutual disposition of the two classes ω_i, ω_j that corresponds to the case of the vectors $\mathbf{V}$ and $\mathbf{m}_{i,j} = (\mathbf{m}_i - \mathbf{m}_j)$ direction coincidence. Herewith, $\alpha = 0$; $tr(\mathbf{V}^T \mathbf{S}_1^{(i,j)} \mathbf{V}) = (\Delta_{i,j})^2$, while parameter $a_{i,j}$ in the area of small values, $\Delta_{i,j}$, can be defined through error function approximation by polynomial function as follows:

$$a_{i,j} \approx \frac{1}{8\sqrt{\pi}x_{i,j}}\left(1 - \frac{x_{i,j}^2}{3} + \frac{x_{i,j}^4}{2!5}\right), \tag{5}$$

where $x_{i,j} = \left(\dfrac{\Delta_{i,j}}{2\sqrt{2}}\right)$, $\Delta_{i,j} \leq \sqrt{2}, x_{i,j} \leq 0,5$.

This method of determining the weighting functions $a_{i,j} = a(\Delta_{i,j})$ can be used to solve the multi-class task, assuming that each c-class has a matrix of intra-group scattering defined by the generalized scatter matrix $\mathbf{S}_2 = \sum_{i=1}^{c} n_i \cdot \Sigma_i$, where Σ_i is the sample covariance matrix of the i-th class. Then it is necessary to find the Euclidian distance between the centers of the corresponding classes $\Delta_{i,j}$ for each class pair $(\omega_i, \omega_j; i, j = 1, \ldots, c; i \neq j)$ in the initial L-dimensional space of the spectral signs and determine the weights $a_{i,j}$ using expression (5). Maximization of the criterion J (2) leads to the procedure of finding eigenvectors $\mathbf{W}_i, i = 1, \ldots, c - 1$ and to the analysis of the object groups distribution within the feature space of reduced dimensions.

4 Application for Dangerous Arrhythmias Recognition

The experimental research was aimed at creating reliable methods to detect ventricular fibrillation, but during this research the possibility was also considered of recognizing such arrhythmia in the early stages of its manifestation, namely, at the time when paroxysmal tachycardia becomes ventricular flutter, which is accompanied, as a rule, with a short period of polytopic (torsades de pointes) tachycardia. In particular, the study was aimed at solving the problem of recognizing 3 classes of dangerous arrhythmias: ω_1 - ventricular fibrillation and ventricular flutter, ω_2 - paroxysmal tachycardia and basic rhythm, represented by different forms of extrasystole, ω_3 - polytopic tachycardia (torsades de pointes).

The application effectiveness of the reviewed procedures was assessed on the basis of results from the experiments performed with real data, including electrocardiographic signal recordings over 20 min. All ECG realizations are obtained from the standard MIT-BIH data base of ECG signals.

An ordered set of 28 spectral features obtained in the frequency domain limited by 15 Hz, using overlapped segments was taken as the initial description of the objects represented by 2-second fragments of electrocardiographic signal [5]. Uncorrelated PSD evaluations are calculated within a spectral window of $\Delta f = 0.976$ Hz, but the spacing along the frequency axis was half of this value, i.e., 0.488 Hz. This case provides the possibility of retaining features of the analyzed signal spectrum shape with relatively stable estimates of the PSD. Distribution of objects of three classes ($c = 3$) represented in a space with dimension $L = 28$ is evaluated in a space with dimension $d = 2$, obtained by the traditional approach (1) using weight functions (2). Comparative analysis of the two distributions shows that application of criterion (2) results in a smoother grouping of the objects on the plane.

During the experiments, dividing functions were created, limits of the solution areas were defined and classification errors were determined, which are the reliability criterion for arrhythmias detection. The result of linear discriminant analysis showed that application of the weight functions can decrease the average classification error from 8.2% to 4.6%, which is an indicator of the effectiveness of this optimization procedure. Analysis of the objects that were situated in the intersection zones of the received solutions has shown that they are controversial in terms of classification, since they can be included in one of the alternative classes. This mainly concerns the intersection zones ω_1 and ω_2 with the intermediate class ω_3. An important result is the unmistakable recognition of classes ω_1 and ω_2, which ensures reliable detection of ventricular fibrillation at the stage of its persistent manifestation.

References

1. Clayton, R.H., Murray, A., Campbell, R.W.F.: Frequency analysis of self–terminating ventricular fibrillation. IEEE Computer Society Press, pp. 705–708 (1994)
2. Clayton, R.H., Campbell, R.W.F., Murray, A.: Time–frequency analysis of human polymorphic ventricular tachycardia. Comput. Cardiol. **24**, 97–100 (1997)
3. Duda, Z., Hart, P.: Pattern classification and Scene Analysis. Mir, Moscow, p. 511 (1976). Translated from English
4. Loog, M., Duin, R.P.W., Haeb-Umbach, R.: Multiclass linear dimension reduction by weighted pairwise Fisher criteria. IEEE Trans. Pattern Anal. Mach. Intell. **23**(7), 762–766 (2001)
5. Manilo, L.A.: Spectral features harmonization by empirical estimates of distances between groups in biosignals classification problems. – Izvestiya vuzov Rossii. Radioelektronika. Vyp. 3, pp. 20–29 (2006)

Information Processing and Applications

An Approach to Web Information Processing

Anatoly Bobkov[1]([✉]), Sergey Gafurov[2], Viktor Krasnoproshin[1], and Herman Vissia[2]

[1] Belarusian State University, Minsk, Republic of Belarus
anatoly.bobkov@gmail.com, krasnoproshin@bsu.by
[2] Byelex Multimedia Products BV, Oud Gastel, The Netherlands
sergey_gafurov@by.byelex.com, h.vissia@byelex.com

Abstract. The paper deals with information extraction from the Internet. Special attention is paid to semantic relations.

Keywords: Information extraction · Semantic patterns · Ontology-based approach

1 Introduction

Nowadays, information and data are stored mainly on the Internet. The Internet opens up tremendous opportunities for information extraction that is gaining much popularity [1, 2]. Currently, information extraction from web documents becomes predominant. Information can come from various sources, e.g. media, blogs, personal experiences, books, newspaper and magazine articles, expert opinions, encyclopedias, web pages, etc.

Information extraction comprises methods, algorithms and techniques for finding the desired, relevant information and for storing it in appropriate form for future use.

The field of information extraction is well suited to various types of business, government and social applications. Diverse information is of great importance for decision making on products, services, events, persons, organizations.

Creation of systems that can effectively extract meaningful information requires overcoming a number of challenges: identification of documents, knowledge domains, specific opinions, opinion holders, events, activities, as well as representation of the obtained results.

The purpose of this paper is to introduce an approach for solving the problem of effective extraction of meaningful, user-oriented information from the web. Semantic patterns approach and an ontology-based approach are proposed as a solution to the problem.

2 Problem Statement and Solution

Numerous models and algorithms are proposed for web information processing and information extraction [3, 4]. But the problem of effective information extraction from texts in a natural language still remains unsolved. Processing of texts in a natural

© Springer International Publishing AG 2017
V.V. Krasnoproshin and S.V. Ablameyko (Eds.): PRIP 2016, CCIS 673, pp. 121–131, 2017.
DOI: 10.1007/978-3-319-54220-1_13

language necessitates the solution of the problem of extracting meaningful information. Semantic relations play a major role [5, 6].

In information extraction and text mining, word collocations show a great potential [7] to be useful in many applications (machine translation, natural language processing, lexicography, word sense disambiguation, etc.).

"Collocations" are usually described as "sequences of lexical items which habitually co-occur, but which are nonetheless fully transparent in the sense that each lexical constituent is also a semantic constituent" [8].

The traditional method of performing automatic collocation extraction is to find a formula based on the statistical quantities of words to calculate a score associated to each word pair. The formulas are mainly: "mutual information", "t-test", "z test", "chi-squared test" and "likelihood ratio" [9].

Word collocations from the point of semantic constituents have not yet been widely studied and used for extracting meaningful information, especially when processing texts in a natural language.

The proposed semantic patterns approach is based on word collocations on the semantic level and contextual relations. Semantic relations (lexical-semantic relations) are meaningful associations between two or more concepts or entities. They can be viewed as links between the concepts or entities that participate in the relation. Associations between concepts can be categorized into different types.

A semantic pattern can be viewed as containing slots that need to be filled. Though most patterns are binary ones having two slots, a pattern may have three or more slots. In general, the proposed semantic patterns include: (1) participants (a person, company, natural/manufactured object, as well as a more abstract entity, such as a plan, policy, etc.) involved in the action or being evaluated; (2) actions - a set of verb semantic groups and verbal nouns ("buy", "build", "arrival"); (3) special-purpose rules representing expert knowledge. The patterns cover different types of semantic relations: (1) semantic relations between two concepts/entities, one of which expresses the performance of an operation or process affecting the other ("Much remains to be learned about how nanoparticles affect the environment"); (2) synonymous relationships ("beautiful – attractive - pretty"); (3) antonymy ("wet - dry"); (4) causal relations ("Research identifies new gene that causes osteoporosis"); (5) hyponymous relations ("Jaguar is a powerful vehicle"); (6) locative relations ("Amsterdam is located in the western Netherlands, in the province of North Holland"); (7) part-whole relations ("car transmission - car"); (8) semantic relations in which a concept indicates a time or period of an event designated by another concept ("Second World War, 1939–1945"); (9) associative relations ("baker – bread": "The baker produced bread of excellent quality"); (10) "made-of" relations ("This ring is made of gold"); (11) "made-from" relations ("Cheese made from raw milk imparts different flavors and texture characteristics to the finished cheese"); (12) "used-for" relations ("Database software is used for the management and storage of data and databases"); (13) homonym relations ("bank of the river – bank as a financial institution"), etc. A semantic relation can be expressed in many syntactic forms. Besides words, semantic relations can occur at higher levels of text (between phrases, clauses, sentences and larger text segments), as well as between documents and sets of documents. The variety of semantic relations and their properties play an important role in web information processing for extracting relevant fragments of information from unstructured text documents.

An ontology-based approach is used for semantic patterns actualization [10].

Ontologies have become common on the World-Wide Web [11]. The broadened interest in ontologies is based on the feature that they provide a machine-processable semantics of information sources that can be communicated among agents as well as between software artifacts and humans. More recently, the notion of ontologies has attracted attention from such fields as intelligent information integration, cooperative information systems, information retrieval, electronic commerce, and knowledge management. For any given knowledge domain, an ontology represents the concepts which are held in common by the participants in a particular domain.

Since ontologies explicitly represent knowledge domain semantics (terms in the domain and relations among them), they can be effectively used in solving information extraction problems, word sense disambiguation in particular.

3   Implementation of the Proposed Approach

The proposed approach has been successfully realized in BuzzTalk portal [12] for subject domains recognition, opinion mining, mood state detection, event extraction, economic activities detection and named entity recognition.

BuzzTalk is offered to companies as a SaaS (Software as a Service) model and it answers questions like:
What are the burning world and local issues?
Who is involved in burning issues?
What are the consequences?
What is the latest information about my competitors?
What are people writing about my product, organization or CEO?
What are important trends in my industry?
What are the big events inside my industry sector?
Where are my customers located?
When and where are people discussing my brand?
BuzzTalk presents a new way of finding content.

The difference between a traditional search engine and a discovery engine such as BuzzTalk, is that search engines list all results for a specific search whereas BuzzTalk allows you to monitor topic-specific developments within your search. BuzzTalk discovers the latest information about a particular brand, competitors or industry, thus facilitating to make better decisions.

BuzzTalk collects all text documents from over 58 000 of the most active websites around the globe, two thirds are news sites and one third are blog sites. The authors of these documents are mainly scientists, journalists and opinion leaders.

BuzzTalk finds and links relevant information in natural-language documents while ignoring extraneous, irrelevant information.

BuzzTalk presents a list of articles in chronological order based on publication date. This list grows each day. You can sort and filter this list based on a variety of criteria such as sentiment, mood state, happenings, etc., thus to experience the wealth of real time information without the pain of information overload. For example, you can easily find all publications within your theme that relate to product releases, employment changes, merger & acquisitions and many more.

Below are examples of information extraction in BuzzTalk.

3.1 Economic Activities Detection

BuzzTalk detects 233 economic activities from texts in a natural language. The economic activities cover all major activities represented in NACE classification (Statistical Classification of Economic Activities in the European Community), which is similar to the International Standard Industrial Classification of all economic activities (ISIC) reflecting the current structure of the world economy. The classifications provide the internationally accepted standard for categorizing units within an economy. Categories of the classifications have become an accepted way of subdividing the overall economy into useful coherent industries that are widely recognized and used in economic analysis, and as such they have become accepted groupings for data used as indicators of economic activities. The classifications are widely used, both nationally and internationally, in classifying economic activity data in the fields of population, production, employment, gross domestic product and others. They are basic tools for studying economic phenomena, fostering international comparability of data and for promoting the development of sound national statistical systems. The classifications provide a comprehensive framework within which economic data can be collected and reported in a format that is designed for purposes of economic analysis, decision-taking and policy-making.

While extracting and analyzing economic activities, BuzzTalk ensures a continuing flow of information that is indispensable for the monitoring, analysis and evaluation of the performance of an economy over time. Moreover, BuzzTalk facilitates information extraction, presentation and analysis at detailed levels of the economy in an internationally comparable, standardized way.

Examples of economic activities detection:

- *Toyota has maintained its position as the world's biggest car manufacturer.*
  Extracted instances:
  Economic activities = ***Manufacture of motor vehicles*** (NACE code C291)
- *The world's first auto show was held in England in 1895.*
  Extracted instances:
  Economic activities = ***Organisation of conventions and trade shows*** (NACE code N823)
- *Goat cheese has been made for thousands of years, and was probably one of the earliest made dairy products.*
  Extracted instances:
  Economic activities = ***Manufacture of dairy products*** (NACE code C105)

- *This invention relates to a process for the hardening of metals.*
  Extracted instances:
  Economic activities = ***Treatment and coating of metals*** (NACE code C256)
- *India is the largest grower of rice.*
  Extracted instances:
  Economic activities = ***Growing of rice*** (NACE code A0112)
- *OCBC Bank operates its commercial banking business in 15 countries.*
  Extracted instances:
  Economic activities = ***Monetary intermediation*** (NACE code K641)
- *It is even more important to properly plan the preparation of legal documents.*
  Extracted instances:
  Economic activities = ***Legal activities*** (NACE code M691)
- *Doran Polygraph Services specializes in professional certified polygraph testing utilizing the latest equipment and most current software with techniques approved by the American Polygraph Association.*
  Extracted instances:
  Economic activities = ***Security and investigation activities*** (NACE code N80)
- *Florida's aquafarmers grow products for food (fish and shellfish).*
  Extracted instances:
  Economic activities = ***Aquaculture*** (NACE code A032)

3.2 Event Extraction

A specific type of knowledge that can be extracted from texts is an event, which can be represented as a complex combination of relations. Event extraction is beneficial for accurate breaking news analysis, risk analysis, monitoring systems, decision making support systems, etc.

BuzzTalk performs real-time extraction of 35 events, based on lexical-semantic patterns, for decision making in different spheres of business, legal and social activities. The events include: "Environmental Issues", "Natural Disaster", "Health Issues", "Energy Issues", "Merger & Acquisition", "Company Reorganization", "Competitive Product/Company", "Money Market", "Product Release", "Bankruptcy", "Bribery & Corruption", "Fraud & Forgery", "Treason", "Hijacking", "Illegal Business", "Sex Abuse", "Conflict", "Conflict Resolution", "Social Life", etc.

Examples:

- *Contract medical research provider, Quintiles, agreed to merge with healthcare information company, IMS Health to make a giant known as Quintiles IMS in an all-stock deal.*
  Extracted instances:
  Event = ***Merger & Acquisition***
- *Mazda Motor Corporation unveiled the all-new Mazda CX-5 crossover SUV.*
  Extracted instances:
  Event = ***Product Release***
- *TCS ranked as top 100 U.S. brand for second consecutive year.*

Extracted instances:
Event = *Competitive Product/Company*
- *Two Hong Kong men arrested for drug trafficking.*
Extracted instances:
Event = *Illegal Business*
- *A former President of Guatemala, already in jail, has been accused of taking bribes.*
Extracted instances:
Event = *Bribery & Corruption*
- *Yet another green-energy giant faces bankruptcy.*
Extracted instances:
Event = *Bankruptcy*
- *Two Afghans held for attempted rape of woman on Paris train.*
Extracted instances:
Event = *Sex Abuse*

The extracted events play a crucial role in daily decisions taken by people of different professions and occupation.

3.3   Subject Domains Recognition

In BuzzTalk a subject domain is recognized on the basis of a particular set of noun and verb phrases unambiguously describing the domain.

Examples:

- *The Forest Inn Hotel offers hotel accommodation on a weekly basis.*
Extracted instances:
Subject domain = *Travel-Hotel*
- *The goal of the pollution prevention and reduction program is to prevent or minimize polluting discharges.*
Extracted instances:
Subject domain = *Ecology*
- *Mozzarella cheese is a sliceable curd cheese originating in Italy.*
Extracted instances:
Subject domain = *Food*
- *Fresh milk is the common type of milk available in the supermarket.*
Extracted instances:
Subject domain = *Beverage*
- *Distance education includes a range of programs, from elementary and high school to graduate studies.*
Extracted instances:
Subject domain = *Education*
- *The biathlon is a winter sport that combines cross-country skiing and rifle shooting.*
Extracted instances:
Subject domain = *Sport*

- *The aim of nanoelectronics is to process, transmit and store information by taking advantage of properties of matter that are distinctly different from macroscopic properties.*
  Extracted instances:
  Subject domain = *Innovation*
- *Britain has made a political decision that will have economic effects.*
  Extracted instances:
  Subject domain = *Politics*
- *Economy from then on meant national economy as a topic for the economic activities of the citizens of a state.*
  Extracted instances:
  Subject domain = *Economics*
- *The law-making power of the state is the governing power of the state.*
  Extracted instances:
  Subject domain = *Law*
- *The president called for collective efforts to fight world terrorism.*
  Extracted instances:
  Subject domain = *Terrorism*
- *Japan was hit by a magnitude 6.5 earthquake followed by an M7.3 quake on Saturday.*
  Extracted instances:
  Subject domain = *Disaster*

For solving the problem of disambiguation special filters, based on the contextual environment (on the level of phrases and the whole text), are introduced.

Subject domains and their concepts are organized hierarchically to state "part-of", "is a kind of" relations.

3.4 Named Entity Recognition

Named-entity recognition is a subtask of information extraction that seeks to locate and classify named entities in a text into pre-defined categories such as names of persons, organizations, locations, etc.

BuzzTalk recognizes the following main named entities:

1. Person (first, middle, last names and nicknames, e.g. Steve Jobs, Cristina Fernandez de Kirchner);
2. Title (social, academic titles, etc.);
3. Position (a post of employment/office/job, e.g. president, CEO);
4. Organization (a company, governmental, military or other organizations, e.g. Microsoft, Wells Fargo, The University of Oxford);
5. Location (names of continents, countries, states, provinces, regions, cities, towns, e.g. Africa, The Netherlands, Amsterdam);
6. Technology (technology names or a description of the technology, e.g. 4D printing, advanced driver assistance, affinity chromatography, agricultural robot, airless tire technology);

7. Product (e.g. Sikorsky CH-148 Cyclone, Lockheed Martin F-35 Lightning II, Kalashnikov AKS, Windhoek Lager, Mercedes S550, Apple iPhone 6S Plus, Ultimate Player Edition, Adenosine);
8. Event (a planned public/social/business occasion, e.g. Olympic Summer Games, World Swimming Championship, Paris Air Show, International Book Fair);
9. Industry Term (a term related to a particular industry, e.g. advertising, finance, aviation, automotive, education, film, food, footwear, railway industries);
10. Medical treatment (terms related to the action or manner of treating a patient medically or surgically, e.g. vitamin therapy, vaccination, treatment of cancer, vascular surgery, open heart surgery)

The named entities are hierarchically structured, thus ensuring high precision and recall.

For example:
Organization

- airline company
- automaker
- bank
- football club
- computer manufacturer
- educational institution
- food manufacturer
- apparel manufacturer
- beverage manufacturer

3.5 Opinion Mining

Opinion mining is gaining much popularity within natural language processing [13]. Web reviews, blogs and public articles provide the most essential information for opinion mining. This information is of great importance for decision making on products, services, persons, events, organizations.

The proposed ontology-based approach for semantic patterns actualization was realized in the developed knowledge base, which contains opinion words expressing:

(1) appreciation (e.g. efficient, stable, ideal, worst, highest);
(2) judgment (e.g. decisive, caring, dedicated, intelligent, negligent)

Opinion words can be expressed by: an adjective (*brilliant, reliable*); a verb (*like, love, hate, blame*); a noun (*garbage, triumph, catastrophe*); a phrase (*easy to use, simple to use*). Adjectives derive almost all disambiguating information from the nouns they modify, and nouns are best disambiguated by directly adjacent adjectives or nouns.

Information about the force of evaluation (low, high, the highest) and orientation (positive/negative) is also included in the knowledge base. For example, *safe* (low force, positive orientation), *safer* (high force, positive orientation), *the safest* (the highest force, positive orientation), *unsafe* (low force, negative orientation).

In the knowledge base opinion words go together with their accompanying words, thus forming "opinion collocations" (e.g. *deep depression*, *deep devotion*, *warm greetings*, *discuss calmly*, *beautifully furnished*). By an "opinion collocation" we understand a combination of an opinion word and accompanying words, which commonly occur together in an opinion-oriented text. The use of opinion collocations is a way to solve the problem of opinion word sense disambiguation (e.g. *well-balanced political leader* and *well-balanced wheel*) and to exclude words that do not relate to opinions (cf. *attractive idea* and *attractive energy*).

We assume that the number of opinion collocations, which can be listed in a knowledge base, is fixed.

The use of opinion collocations within the ontology-based approach opens a possibility to assign names of knowledge domains to them, because opinion collocations are generally domain specific. For example, *helpful medical staff* ("health care"), *helpful hotel reception staff* ("travel-hotel"), *stable economy* ("economics"), *well-balanced politician* ("politics"). More than one knowledge domain may be assigned to an opinion collocation, e.g. *fast service* ("economics-company", "travel-hotel").

Processing of the extracted opinion collocations is carried out in their contextual environment. The developed algorithm checks for the presence of modifiers that can change the force of evaluation and orientation indicated in the knowledge base.

The developed knowledge base also provides additional information about quality characteristics and relationships for different objects on which an opinion is expressed (e.g. *software product* evaluation includes: usability, reliability, efficiency, reusability, maintainability, portability, testability; *travel-hotel* evaluation includes: value, rooms, location, cleanliness, check in/front desk, service).

The results of opinion collocations processing are grouped and evaluated to recognize the quality of the opinion-related text. The results are also visualized.

3.6 Mood State Detection

A valuable addition to opinion mining is detection of individual/public mood states. The relationship between mood states and different human activities has proven a popular area of research [14].

BuzzTalk mood detection uses the classification of the widely-accepted "Profile of Mood States" (POMS), originally developed by McNair, Lorr and Droppleman [15].

In BuzzTalk, mood state detection is based on: (1) mood indicators (e.g. "I feel", "makes me feel", etc.); (2) mood words (e.g. anger, fury, horrified, tired, taken aback, depressed, optimistic); (3) special contextual rules to avoid ambiguation. BuzzTalk automatically recognizes the following mood states: "Anger", "Tension", "Fatigue", "Confusion", "Depression", "Vigor".

Examples:

- *Despite these problems, I feel very happy.*
  Extracted instances:
  Mood state = ***Vigor***
- *I'm feeling angry at the world now.*

Extracted instances:
Mood state = *Anger*
- *I feel fatigued and exhausted.*
Extracted instances:
Mood state = *Fatigue*
- *I have suicidal thoughts every day.*
Extracted instances:
Mood state = *Depression*

Mood state detection alongside with opinion mining can give answers to where we are now and where will be in future.

4 Conclusion

With the rapid growth of the Internet there is an ever-growing need for reliable multi-functional systems to retrieve relevant and valuable information.

The proposed semantic patterns approach has been successfully realized in BuzzTalk portal for opinion mining, mood state detection, named entity recognition, economic activities detection, subject domain recognition, event extraction. It plays a vital role in information extraction and new knowledge discovery from web documents. The approach ensures high accuracy, flexibility for customization and future diverse applications for information extraction. New semantic relations can be easily created. The relations can be decomposed into simpler relational elements. Semantic relations follow certain general patterns and rules, the same types of semantic relations are used in different languages.

Semantic word collocations are a major factor in the development of a wide variety of applications including information extraction and information management (retrieval, clustering, categorization, etc.).

Implementation results show that the proposed knowledge-based approach is correct and justified and the technique is highly effective.

References

1. Moens, M.: Information Extraction: Algorithms and Prospects in a Retrieval Context, p. 246. Springer, Amsterdam (2006)
2. Baeza-Yates, R., Ribeiro-Neto, B.: Modern Information Retrieval: The Concepts and Technology behind Search, p. 944. Addison-Wesley Professional, Harlow (2011)
3. Buettcher, S., Clarke, C., Cormack, G.: Information Retrieval: Implementing and Evaluating Search Engines, p. 632. MIT Press, Cambridge (2010)
4. Machová, K., Bednár, P., Mach, M.: Various approaches to web information processing. Comput. Inf. **26**, 301–327 (2007)
5. Khoo, Ch., Myaeng, S.H.: Identifying semantic relations in text for information retrieval and information extraction. In: Green, R., Bean, C.A., Myaeng, S.H. (eds.) The Semantics of Relationships: An Interdisciplinary Perspective, pp. 161–180. Springer, Amsterdam (2002)

6. Bobkov, A., Gafurov, S., Krasnoproshin, V., Romanchik, V., Vissia, H.: Information extraction based on semantic patterns. In: Proceedings of the 12th International Conference – PRIP 2014, Minsk, pp. 30–35 (2014)
7. Barnbrook, G., Mason, O., Krishnamurthy, R.: Collocation: Applications and Implications, p. 254. Palgrave Macmillan, Basingstoke (2013)
8. Cruse, D.A.: Lexical Semantics, p. 310. Cambridge University Press, Cambridge (1986)
9. Manning, C.D., Schütze, H.: Foundations of statistical natural language processing, p. 620. MIT Press, Cambridge (1999)
10. Bilan, V., Bobkov, A., Gafurov, S., Krasnoproshin, V., van de Laar J., Vissia, H.: An ontology-based approach to opinion mining. In: Proceedings of 10th International Conference PRIP 2009, Minsk, pp. 257–259 (2009)
11. Fensel, D.: Foundations for the Web of Information and Services: A Review of 20 Years of Semantic Web Research, p. 416. Springer, Heidelberg (2011)
12. http://www.buzztalkmonitor.com
13. Pang, B., Lee, L.: Opinion Mining and Sentiment Analysis, p. 148. Now Publishers Inc., Hanover (2008)
14. Clark, A.V.: Mood State and Health, p. 213. Nova Publishers, Hauppauge (2005)
15. McNair, D.M., Lorr, M., Droppleman, L.F.: Profile of Mood States. Educational and Industrial Testing Service, San Diego (1971)

Fuzzy Semi-supervised Clustering with Active Constraint Selection

Natalia Novoselova(✉) and Igor Tom

United Institute of Informatics Problems, NAS Belarus, Surganova Street 6,
220012 Minsk, Belarus
{novosel,tom}@newman.bas-net.by

Abstract. The paper presents the approach to semi-supervised fuzzy clustering, based on the extended optimization function and the algorithm of the active constraints selection. The approach is tested on the artificial and real data sets. Clustering results, obtained by the proposed approach, are more accurate relative to the ground truth due to utilization of the additional information about the class labels in the most uncertain regions.

Keywords: Semi-supervised clustering · Pairwise constraints · Active selection

1 Introduction

In the field of machine learning and bioinformatics the semi-supervised methods present the realization of the technology which uses the data with both known and unknown class labels in order to solve some particular task, such as data clustering or classification. As a rule the number of unlabeled data considerably exceeds the number of the data with known labels. It can be explained by high financial and temporal expenses, connected with the manual data classification in such fields of research as natural language processing, text mining, computational biology etc. In order to make use of a tremendous amount of rapidly coming information, which due to the progress in information technologies can be unlimitedly stored in data bases the special methods of semi-supervised learning are currently of great concern. It is recognized that the unlabeled data together with the sufficiently small amount of constrained ones enable the significant improvement in learning accuracy [1].

The semi-supervised clustering is one of the evolving research directions, used for the data exploratory analysis. The main task of the clustering methods is to reveal the groups of similar data points, according to the specified notion of similarity. There are already several approaches to consider the known constraints between the points in order to guide the clustering process [2, 3]. The semi-supervised learning allows improving the efficiency of the clustering using the available expert knowledge in the form of data labels or the relations between the data points.

For the semi-supervised algorithm, processing the great amount of data, as e.g. biological data it is very important to:

(1) automatically determine the number of clusters in the data;
(2) take into consideration the available data constraints;

V.V. Krasnoproshin and S.V. Ablameyko (Eds.): PRIP 2016, CCIS 673, pp. 132–139, 2017.
DOI: 10.1007/978-3-319-54220-1_14

(3) automatically select the constraints in more uncertain, transition regions in order to get the high accuracy of results. It must be emphasized that the defined constraints greatly influence the clustering result. The improper constraint selection can even decrease the clustering performance [4]. Recently the most important topic of research is the development of active selection strategies, which search for the most useful constraints [5, 6]. They can minimize the expenses of getting the labeled information without loss of clustering accuracy.

In the paper we propose the approach to semi-supervised fuzzy clustering, based on the active constraint selection algorithm [7]. The semi-supervised fuzzy clustering algorithm [8] takes into account the pairwise constraints and belongs to the class of optimization clustering methods. The basis of such methods is the construction of some optimization function the minimization of which enables to define the optimal cluster parameter values. The experiments on several datasets have shown the improvement of clustering performance with the inclusion of constraints, especially when they were actively selected.

2 Semi-supervised Fuzzy Clustering

In our research we have adopted the fuzzy clustering algorithm, proposed in [8], which is based on the algorithm of competitive agglomeration. The algorithm can automatically determine the number of clusters in the analyzed data and takes into account the data constraints using the extended clustering optimization function. There are two types of constraints: "must link" constraint and "cannot link" constraint for data points. Let M is the set of "must-link" constraints, i.e. $(x_i, x_j) \in M$ means the data points x_i and x_j lie in the same cluster. The set Q consists of "cannot-link" constraints, i.e. $(x_i, x_j) \in Q$ means the data points x_i and x_j lie in the different clusters. The extended optimization function is the following

$$
J(V, U) = \sum_{k=1}^{C} \sum_{i=1}^{N} (u_{ik})^2 d^2(x_i, \mu_k) - \alpha \left(\sum_{(x_i,x_j) \in M} \sum_{k=1}^{C} \sum_{l=1,l \neq k}^{C} u_{ik} u_{jl} + \right.
$$
$$
\left. \sum_{(x_i,x_j) \in Q} \sum_{k=1}^{C} u_{ik} u_{jk} \right) - \beta \sum_{k=1}^{C} \sum_{i=1}^{N} (u_{ik})^2
$$
(1)

where $X = \{x_i | i \in \{1, \dots, N\}\}$ is the dataset of size N, $V = \{\mu_k | k \in \{1, \dots, C\}\}$ is the centers of C clusters, $U = \{u_{ik} | k \in \{1, \dots, C\}, i \in \{1, \dots, N\}\}$ is the set of membership degrees. The constraint $\sum_{k=1}^{C} u_{ik} = 1, i = \{1, \cdots, N\}$ must be considered.

The cluster centers $(1 \leq k \leq C)$ are calculated in iterative fashion as

$$\mu_k = \frac{\sum_{i=1}^{N} (u_{ik})^2 x_i}{\sum_{i=1}^{N} (u_{ik})^2} \tag{2}$$

and the cardinalities of the clusters are defined as

$$N_s = \sum_{i=1}^{N} u_{is}. \tag{3}$$

The first component of optimization function (1) presents the FCM optimization term and considers the cluster compactness. The second component consists of two terms: (1) penalty for the violation of the pairwise "must-link" constraints; (2) penalty for the violation of the pairwise "cannot-link" constraints. The weight constant α determines the relative importance of supervision. The third component in (1) is the sum of squares of cardinalities of the individual clusters and corresponds to the regularization term, which controls the number of clusters. The weight function β of the third component provides the balance between the components and is expressed as

$$\beta(t) = \frac{\eta_0 exp(-|t - t_0|/\tau)}{\sum_{j=1}^{C} \left(\sum_{i=1}^{N} u_{ij} \right)^2} \times \left[\sum_{j=1}^{C} \sum_{i=1}^{N} u_{ij}^2 d^2 (x_j, \mu_j) \right] \tag{4}$$

Function $\beta(t)$ allows regulating the data memberships u_{ij} to clusters and has the small value at the beginning of the optimization process in order to form the initial clusters. After that the weight rises in order to reduce the number of clusters and again falls to diminish its influence on the cluster formation.

We have reconsidered the derivation of the expressions for the modification of the parameters $u_{rs} = u_{rs}^{FCM} + u_{rs}^{constr} + u_{rs}^{bias}, r = \{1, \cdots, N\}, s = \{1, \cdots, C\}$ in (1). The expressions are calculated using the Lagrange multipliers and are the following:

$$
\begin{aligned}
u_{rs}^{FCM} &= \frac{1}{d^2 (x_r, \mu_s)} \Big/ \sum_{k=1}^{C} \frac{1}{d^2 (x_r, \mu_k)} \\
u_{rs}^{constr} &= \frac{\alpha}{2d^2 (x_r, \mu_s)} \left(\overline{C_{v_r}} - C_{v_{rs}} \right) \\
u_{rs}^{bias} &= \frac{\beta}{d^2 (x_r, \mu_s)} \left(N_s - \overline{N_r} \right)
\end{aligned} \tag{5}
$$

where $C_{v_{rs}}$ – penalty expression for violation the constraint for the rth point in the case of sth cluster, $\overline{C_{v_r}}$ – weighted average penalty for all clusters for the rth point, $\overline{N_r}$ – weighted average of cluster cardinalities relative to rth point.

The term u_{rs}^{FCM} is the same as in FCM; the term u_{rs}^{constr} allows decreasing or increasing the membership according to the pairwise constraints, defined by user; the term u_{rs}^{bias}

allows to reduce the cardinalities of the non-informative clusters and to discard them from consideration when the cluster cardinalities are below the threshold.

The important step of the semi-supervised clustering process is the cluster merging, which is executed at each iteration of the optimization procedure. It allows excluding from consideration not only the small clusters, but also the non-informative clusters of different sizes.

Below is the scheme of the semi-supervised algorithm [8].

Algorithm

* Define the maximal cluster number C.
* Initialize the cluster centers randomly.
* Initialize the membership values of data objects to clusters: the equal membership to each cluster.
* Calculate the initial cardinalities of each cluster.

Repeat

* Calculate β using expression (4).
* Calculate memberships u_{ij} using (5).
* Calculate the cluster cardinalities $N_j, 1 \leq j \leq C$ using (3).
* For each cluster if $N_j <$ threshold   discard cluster j.
* Update number of clusters C.
     Repeat
     * Merge the nearest clusters using the special procedure.
       Until further merging is required
* Update the cluster centers using expression (2).
     Until the clusters stabilize.

3   Active Constraint Selection

In [8] the available constraints, which are selected randomly, significantly increase the performance of data clustering. In our paper we propose to use the active constraint selection algorithm [7], which is able to direct the search for the constraints to the most uncertain (transition) clustering regions. The candidate subset of constraints is constructed on the basis of k-Nearest Neighbor Graph (k-NNG). After that the selection is performed from the list of candidate constraints, sorted according to their ability to separate clusters. As a rule such constraints lie in the most uncertain clustering regions.

The k-NNG graph is constructed using the information about k- nearest neighbors of each data point. The weight $w(x_i, x_j)$ of the graph edge between two points x_i and x_j is defined as the number of their common nearest neighbors

$$w(x_i, x_j) = \left| NN(x_i) \cap NN(x_j) \right|, \tag{6}$$

where $NN(x)$ is the k- nearest neighbors of point x.

The constraint ability to separate clusters is estimated using the following utility measure

$$ASC(x_i, x_j) = \frac{k - w(x_i, x_j) + \dfrac{1}{1 + \min\{LDS(x_i),\, LDS(x_j)\}}}{k+1}, \tag{7}$$

where $LDS(x) = \dfrac{\sum_{q \in NN(x)} w(x, q)}{k}$ is the local density of point x. The ASC measure of pairwise constraint (x_i, x_j) depends on the corresponding edge weight $w(x_i, x_j)$ and the constraint density, which is defined by minimum of local densities of points x_i and x_j. The ASC measure helps to reveal the constraints, which are more informative for clustering, i.e. can improve the clustering performance. The higher ASC value corresponds to more informative constraint.

The candidate constraints are selected using the k-NNG graph as follows

$$C = \{(u, v) \mid w(u, v) < \theta\}, \tag{8}$$

where u, v – graph vertices, θ – the threshold parameters, defined in the interval $\left[\dfrac{k}{2} - 2, \dfrac{k}{2} + 2\right]$.

In order to refine the constraint selection process the authors in [7] propagate the already selected constraints to the whole set of candidate constraints. The propagation procedure helps to exclude from further consideration the constraints, which can be derived from the already selected ones using the strong path and transitive closure concepts.

According to the algorithm [7] the constraints are defined iteratively, starting from the one constraint till the required number. Each constraint can be selected from the candidate subset C using two variants: (1) random choice of constraints; (2) taking from the constraint list, sorted according to ASC measure. In our research we have compared the active constraint selection procedure, based on ASC measure with purely random choice of constraints from the data.

4   Results of Experiments

Several comparative experiments using fuzzy semi-supervised clustering algorithm with active constraints (AS) and with purely random constraints (RS) were conducted on artificial and real data sets. The results were compared with the simple k-means (KM) and competitive agglomeration algorithm (CA).

The artificial data set Data1 consists of 150 objects with two features. The objects are divided into three clusters, generated according to the multivariate normal distribution and are partly overlapped. The real data set Leukemia consists of two classes – 47 samples of acute lymphoblastic leukemia (ALL) and 25 samples of acute myeloid leukemia (AML) [9]. In order to validate our approach we have taken into account the two subtypes of ALL: 38 samples of B-cell ALL and 9 samples of T-cell T-ALL, analyzing the classification into 3 classes (ground truth). All the samples are characterized by the expression of 7129 genes. After data preprocessing with thresholding and filtering the 3571 genes are selected for further analysis.

The clustering quality was estimated with the external validation criterion using the ground truth. The criterion estimates the similarity of two data partitions. The first partition corresponds to the known class labels. The second partition is calculated on the basis of fuzzy clustering results, where the label for each data point corresponds to the cluster to which it has the highest membership. As the labels of data points in two partitions can be permuted it is necessary to find the correspondence between them, solving the following optimization task:

Let α_1 and α_2 are two class label functions, defined by two partitions $\Pi_k^{(1)}$, $\Pi_k^{(2)}$ of the set X into k groups, i.e. $\alpha_i(x) = j$, if and only if $x \in \pi_j^{(i)}$, $i = 1, 2\ j = 1, \ldots, k$. For given permutation φ of class labels from set V_k consider the empirical validation criterion:

$$d_k(\alpha_1, \alpha_2, \varphi) = \frac{1}{|X|} \sum_{x \in X} \alpha_1(x) \neq \varphi(\alpha_2(x)), \tag{9}$$

where δ – indicator function $\alpha_1(x) \neq \varphi(\alpha_2(x))$:

$$\delta(\alpha_1(x) \neq \varphi(\alpha_2(x))) = \begin{cases} 1, & \text{if } \alpha_1(x) \neq \varphi(\alpha_2(x)) \\ 0, & \text{otherwise} \end{cases}. \tag{10}$$

The optimal class label permutation φ^* is defined as

$$\varphi^* = \arg\min_{\varphi} d_k(\alpha_1, \alpha_2, \varphi) \tag{11}$$

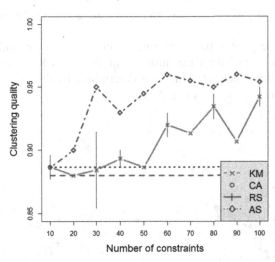

Fig. 1. Clustering results for the artificial dataset using cluster validation criterion

Figures 1 and 2 present the dependence between the validation criterion and the number of pairwise constraints considered for AS and RS algorithms. The number of constraints is in the range from 0 to 100 in increments of 10. For every number of constraints, 100 experiments were performed with different random selections of the constraints in order to estimate the standard errors for the RS approach. KM and CA algorithms don't consider the constraints and are depicted for reference. The algorithms CA, RS and AS were initialized with more than real number of clusters and the search for real clustering structure is performed automatically during algorithm execution. According to Figs. 1 and 2 including the random constraints into clustering allows to improve the clustering quality.

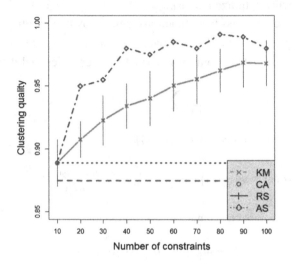

Fig. 2. Clustering results for Leukemia dataset using cluster validation criterion

AS algorithm improves the clustering results even more and requires fewer constraints in order to reach the same clustering quality as the RS algorithm. The cluster centers for the artificial dataset, which are determined by the fuzzy semi-supervised clustering algorithm are shown in Fig. 3.

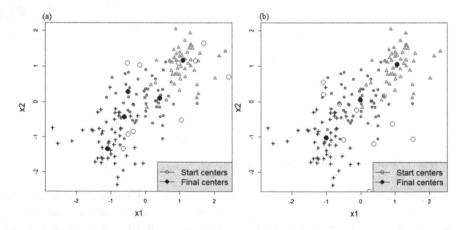

Fig. 3. Initial and final cluster centers for artificial dataset: a) clustering with 10 random constraints; b) clustering with 10 active constraints.

According to Fig. 3 the application of active constraints not only helps to raise the cluster validation measure but also to improve the search for the real number of clusters.

5 Conclusion

The paper presents the approach to semi-supervised fuzzy clustering with active constraints selection. The extended clustering optimization function of the clustering algorithm takes into account the "must link" and "cannot link" constraints on pairwise data positions in the clusters and is based on the scheme, proposed in [8]. We have applied the algorithm of the active constraints selection [7] to generate the constraints for experimental datasets. The clustering results have shown the improved performance with both random and active constraints, included into fuzzy clustering process. The inclusion of active constraints led to better clustering results and to less number of constraints to attain the high level of the clustering quality. Moreover the active constraints help to define the real number of clusters in the competitive agglomeration process. The algorithms' realization, data modeling and experiments were performed in R Studio environment using the R language [10].

References

1. Chapelle, O., Schölkopf, B., Zien, A.: Semi-Supervised Learning. MIT Press, Cambridge, MA, USA (2006). ISBN 0-262-25589-8
2. Basu, S., Banerjee, A., Mooney, R.J.: Semi-supervised clustering by seeding. In: 19th International Conference on Machine Learning (ICML-2002), pp. 19–26. Morgan Kaufmann Publishers Inc., San Francisco (2002)
3. Basu, S., Davidson, I., Wagstaff, K.: Constrained Clustering: Advances in Algorithms, Theory, and Applications. Chapman & Hall, Boca Raton (2008)
4. Wagstaff, K.L.: Value, cost, and sharing: open issues in constrainted clustering. In: 5th International Workshop on Knowledge Discovery in Inductive Databases, pp. 1–10 (2007)
5. Mallapragada, P.K., Jin, R., Jain, A.K.: Active query selection for semi-supervised clustering. In: 19th International Conference on Pattern Recognition (2008). doi:10.1109/ICPR. 2008.4761792
6. Sk, J.A., Prasad, M., Gubbi, A., Rahman, H.: Active Learning of constraints using incremental approach in semi-supervised clustering. Int. J. Comput. Sci. Inf. Technol. **6**(2), 1962–1964 (2015)
7. Vu, V.V., Labroche, N., Bouchon-Meunier, B.: Boosting clustering by active constraint selection. In: 19th European Conference on Artificial Intelligence, ECAI-2010, pp. 297–302 (2010)
8. Grira, N., Crucianu, M., Boujemaa, N.: Semi-supervised fuzzy clustering with pairwise-constrained competitive agglomeration. In: 14th IEEE International Conference on Fuzzy Systems (Fuzz'IEEE 2005), May 2005. doi:10.1109/FUZZY.2005.1452508
9. Golub, T.R., Slonim, D.K., Tamayo, P., Huard, C., Gaasenbeek, M., Mesirov, J.P., Coller, H., Loh, M.L., Downing, J.R., Caligiuri, M.A., Bloomfield, C.D., Lander, E.S.: Molecular classification of Cancer: class discovery and class prediction by gene expression monitoring. Science **286**(5439), 531–537 (1999)
10. R Core Team R: A language and environment for statistical computing. R Foundation for Statistical Computing, Vienna, Austria (2013). http://www.R-project.org

Unmanned Aerial Vehicle (UAV): Back to Base Without Satellite Navigation

Vladislav Blazhko[1(✉)], Alexander Kalinovsky[2], and Vassili Kovalev[2]

[1] Belarusian State University, Minsk, Belarus
mr_plum@mail.ru
[2] United Institute of Informatics Problems of the NAS of Belarus, Minsk, Belarus
gakarak@gmail.com, vassili.kovalev@gmail.com

Abstract. Everyone wants to automate routine or tiring work. There are a lot of tasks that may be done by drones. For example border protection or delivering. But before any company or even country adopts any technology we need to verify that it's not vulnerable to any attacks. A satellite navigation is at least one vulnerability for drones, which can be easily broken or spoofed. This is a serious problem on the way to automation work by drones. Our goal is to improve accuracy of existing Inertial Navigation Systems on UAVs with an on-board video camera. In this article we've investigated how feature based methods fit back to base problem.

Keywords: UAV · Drones · Detectors · Descriptors

1 Introduction

Nowadays UAV's industry evolves in the different spheres of life. Think about the Amazon Prime Air that needs to deliver purchases to customers or those various military drones that can strike or scout, drones for mapping and etc.

The most of the drones have auto-return home function: when satellite navigation is available, drone remembers the exact spot which it took off from. Wherever drone is flying, if its battery is running very low or user decided to call it back then it returns directly to home.

This function works well under normal conditions. But under military conditions, where it is very simple to drown out the GPS signal or even emit counterfeit signal, we are forced to find another methods to return to home. Also in some regions the GPS signal can be unstable and weak and cause problems when returning home.

1.1 Inertial Navigation System

Usually drone has Inertial Navigation System (INS) [1], which contain Inertial Measurement Units (IMU) which have angular and linear accelerometers (for changes in position); some IMUs include a gyroscopic element (for maintaining an absolute angular reference). The purpose of INS is calculate via dead reckoning the position, orientation,

© Springer International Publishing AG 2017
V.V. Krasnoproshin and S.V. Ablameyko (Eds.): PRIP 2016, CCIS 673, pp. 140–149, 2017.
DOI: 10.1007/978-3-319-54220-1_15

and velocity (direction and speed of movement) of a moving object without the need for external references.

The main disadvantage of INS is that they have measurement errors and accumulate them during the flight. Moreover there is a whole class of small drones that cannot afford to equip large or heavy, or expensive INS, it also reduces the accuracy of the measurements.

So if UAV is very far from home and will return only by INS, then it can return to the place which is still far from home. Figure 1 shows accuracy of different navigation systems.

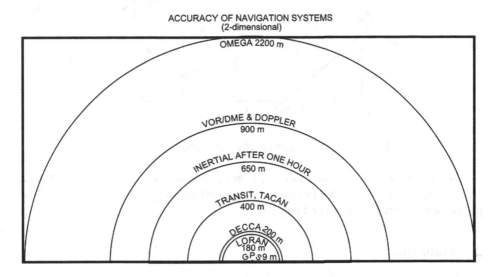

Fig. 1. Accuracy of navigation systems

1.2 Experimental Results with INS

We conducted several tests with INS on real drone AR.Drone 2.0 (Fig. 2). It has weight about 400 grams and flight time about 15 min. Also it has gyroscope with 3 axles, accuracy of 2,000° per second, and accelerometer with 3 axles, accuracy of ± 50 mg. You can find more details at [2].

Fig. 2. AR.Drone 2.0

Figure 3 represents obtained trajectory. The rectangle denotes the room. The drone was flying from the first corner to the second and then from the second to the third corner and then back on the same way.

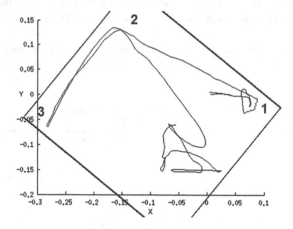

Fig. 3. Trajectory recovered by Inertial Navigate System

As you can see initially the trajectory is good, but over time it gets worse. So we want to clarify the work of INS by developing a small lightweight module, like a mobile phone, with on board-camera and processor.

2   Main Idea

The main idea is very simple. While the drone has a signal from satellite navigation system it will take photos of the area on which it flies with some specific period of time. Each photo is linked to physical coordinates. Thus, the drone can represent photos as a trajectory (Fig. 4).

As soon as the signal of satellite navigation is lost, the drone rises above to easily find the last area on which it flew. And then drone follows the trajectory while returning back.

The *find last area* is an attempt to find the relevant part of catched trajectory on the current camera view (Fig. 5). If we do this then we can calculate translate, rotate and scale transformation. After we can get approximated coordinates of our location, since the part of trajectory linked to physical coordinates. And finally choose the direction of flight.

Fig. 4. Trajectory built on photos linked to coordinates

Fig. 5. Example of the camera view with found a part of the trajectory

The trajectory may be very long and finding the relevant part of whole trajectory on the current camera view can require huge computation. Therefore we will use telemetry from INS to define an approximated scope of the trajectory in which we are. It allow us to find the relevant part on the part of trajectory instead of whole trajectory, which in turn reduces the computational complexity.

3   Computer Vision Methods

The main aim is getting the transformation between current camera view and the relevant part of the trajectory.

More formally the transformation is a matrix of homography, which more correctly as possible transform the points from trajectory to the current camera view. To obtain this matrix we must have a 2D to 2D point correspondences.

So, as a first step, we researched how is good the feature based methods for getting correct transformations. For this purpose we used library OpenCV [3].

Algorithm for getting final transformation presented in Fig. 6. We tested follow feature detectors: SIFT, SURF, KAZE, AKAZE, BRISK, ORB, MSER [4–10]. All detectors have optimal parameters.

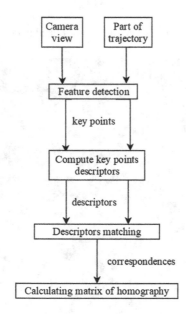

Fig. 6. Algorithm which obtain transformation

The descriptors same like detectors plus FREAK [11]. The matcher used BFMatcher (Brute-Force matcher), because we need quality and want avoid any distortions that may be caused with other matchers. The matcher takes *NORM_L2_SQR* [12] for SIFT, SURF, KAZE and AKAZE descriptors and NORM_HAMMING for BRISK, FREAK and ORB descriptors and k equal to 2 (it means, that BFMatcher will search for 2 near neighbors).

After the matches found, they ratio tested. Ratio test consists in that matches, where the distance between the first and second neighbor point large, are good, because in this case we without doubts can choose between them, otherwise we exclude that match like being able to cause confusion.

For calculating matrix of homography used the function findHomograpy with RANSAC method [13], because not all of the correspondence points fit the rigid perspective transformation.

4 Materials

For our problem we need an algorithm, which will be scale and rotate invariant and fast. So we prepare the dataset, which consists of 4 classes: "no_scale", "scale_1.1", "scale_1.5", "scale_2.0".

Each class contain the main image, which we call map, 49 rotated and scaled (according to the name of class) pieces of this map, which we call frames, and 49 matrixes of homography, which perform transformation from map to frame.

The map obtained from the program SASPlanet [14], which allow to get the photos from satellite with different zoom like an image. The region of interest was selected and downloaded the map of Yandex satellite maps (Fig. 7) and Google satellite maps on zoom 17 (~ 1.41 meters per pixel). This region was selected, because it has many different terrain types: forest, river, buildings, roads and wastelands.

Fig. 7. Map from the Yandex satellite maps

Fig. 8. Example of image from class "scale_2.0"

The map size is 1126×1226 px and the size of each frame is 353×353 px. Example of frame from class "scale_2.0" in Fig. 8. The scaling was done artificially by stretching. Each matrix of homography exactly transform the points of the map to points of the frame, because each matrix was built with known center coordinates, rotate angle and scale multiplier.

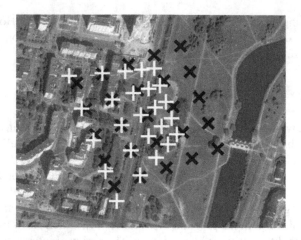

Fig. 9. Example of grid score. Score equal to 0.2

5   Score Discussing

The accuracy of approximated coordinates of drone location directly depend on the accuracy of transformed points.

Grid score. We inflict the uniform grid of N x M points on the frame. And then transform this points on the map by inverse original matrix of homography and inverse of matrix given by algorithm. Then just counting percent of correct points. Point treated as correct if Euclidean distance between corresponding points of original and algorithm transforms on map less or equal than some threshold θ. Thus, the score is a percent of correct points.

$$correct\left(pt_{algo}, pr_{orig}\right) = \begin{cases} 1, & if\ dist\left(pt_{algo}, pt_{orig}\right) < \theta \\ 0, & otherwise \end{cases} \tag{1}$$

Example of grid score in Fig. 9, where the grid is 5×5 points, the white points belong to the grid of algorithm and black - original. There are 5 points match with threshold 5 pixels.

Key points score. This score is approximated score from [15]. We got the matched key points on the map and on the frame. So we transform matched key points on the map to the frame by the original matrix of homography. The score is a percent of correct points as in Eq. (1).

6   Experimental Results

In our experiment we took the grid 10×10 points with threshold equal to 5 pixels. The map from Yandex satellite maps and frames of map from the same map. Also we have experimented with the map from Google satellite maps and frames from Yandex satellite maps, but all algorithms have given almost zero results.

The average time required for detecting and building descriptor for one key point shown in Fig. 10 (less better). The algorithm configuration denoted as detector + descriptor, if there is no plus sign, then detector and descriptor has same name.

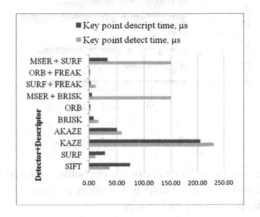

Fig. 10. Plot of time required by detector and descriptor

Mean grid scores for each class of dataset presented in Fig. 11 (more better). Plot consists of 4 layouts, each layout represent the value of grid score for appropriate class. The same plot presented in Fig. 12, but of mean key points score.

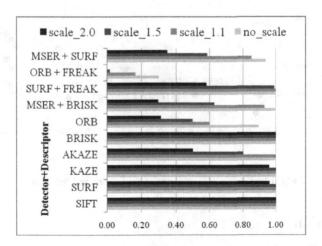

Fig. 11. Mean grid scores

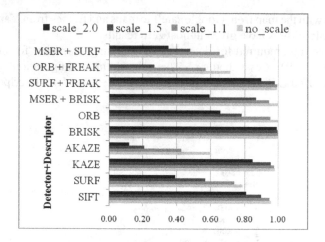

Fig. 12. Mean key points scores

The link between average full time of calculating homography and average grid score shown in Fig. 13. This plot doesn't display the point corresponding to the ORB with FREAK, because this bunch gave bad results.

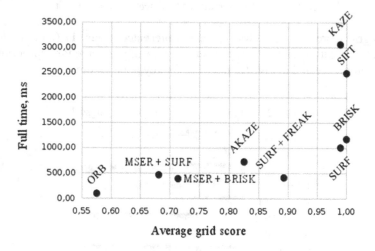

Fig. 13. Scatter plot of average grid score and full time

7   Conclusion

As can be seen from the experiment with map from Google satellite maps and frames from Yandex satellite maps, the approach for accuracy improvement of existing INS can be applied to the drones, which have flight time about several hours. The cause of bad results in this experiment is that the photos from Google and Yandex satellites taken in different seasons and feature detectors select different key points at the same areas. Although it's may be a fundamental problem of feature based methods.

Another conclusion that can be drawn is that the algorithm, which rely on BRISK detector and descriptor, has a good recovery performance of matrix of homography in both scores and at the same time it has a good time of calculating. Also SURF with FREAK has a good performance in both scores and very little time and can compete ORB.

References

1. Kayton, M.: Navigation — Land, Sea, Air, and Space. IEEE Press, New York (1989)
2. AR.Drone 2.0. Parrot new wi-fi quadricopter - AR.Drone.com - HD Camera - Civil drone – Parrot. http://ardrone2.parrot.com/. Accessed 13 Apr 2016
3. OpenCV I OpenCV. http://opencv.org/. Accessed 13 Apr 2016
4. Lowe, D.: Distinctive image features from scale-invariant keypoints. Int. J. Comput. Vis. 2(60), 91–110 (2004)
5. Bay, H., Ess, A., Tuytelaars, T., Gool, L.J.V.: Speeded-up robust features (surf). CVIU 110(3), 346–359 (2008)
6. Alcantarilla, P.F., Bartoli, A., Davison, A.J.: KAZE Features. In: Fitzgibbon, A., Lazebnik, S., Perona, P., Sato, Y., Schmid, C. (eds.) ECCV 2012. LNCS, vol. 7577, pp. 214–227. Springer, Heidelberg (2012). doi:10.1007/978-3-642-33783-3_16
7. Alcantarilla, P.F., Nuevo, J., Bartoli, A.: Fast explicit diffusion for accelerated features in nonlinear scale spaces. In: British Machine Vision Conference (2013)
8. Leutenegger, S., Chli, M., Siegwart, R.:. Brisk: binary robust invariant scalable keypoints. In: ICCV (2011)
9. Rublee, E., Rabaud, V., Konolige, K., Bradski, G.R.: Orb: an efficient alternative to sift or surf. In: ICCV (2011)
10. Matas, J., Chum, O., Urban, M., Pajdla, T.: Robust wide baseline stereo from maximally stable extremal regions. In: Proceedings of British Machine Vision Conference, pp. 384–396 (2002)
11. Alahi, A., Ortiz, R., Vandergheynst, P.: Freak: fast retina keypoint. In: CVPR (2012)
12. Arandjelovic, R., Zisserman, A.: Three things everyone should know to improve object retrieval. In: CVPR (2012)
13. Fischler, M.A., Bolles, R.C.: Random sample consensus: a paradigm for model fitting with applications to image analysis and automated cartography. Commun. ACM 24(6), 381–395 (1981)
14. SASGIS - Веб-картография и навигация >> SAS.Планета. http://sasgis.ru/sasplaneta/. Accessed 14 Apr 2016
15. Mikolajczyk, K., Schmid, C.: A performance evaluation of local descriptors. PAMI 27, 1615–1630 (2004)

Models and Technology for Medical Diagnostics

Vladimir A. Obraztsov[✉] and Olga V. Shut

Belarusian State University, 4 Independence Avenue, 220030 Minsk, Belarus
`obraztsov@bsu.by, olgashut@tut.by`

Abstract. This paper presents a novel decision support approach for the medical diagnostics in sports traumatology. In particular, the corresponding pattern recognition problem is solved by means of a special hybrid algorithm.

Keywords: Pattern recognition · Decision support · Medical diagnostics

1 Introduction

The decision support computer systems are widely used in the modern medical diagnostics. The specific implementation of such systems depends on the medical problem being solved as well as the core mathematical models and computer technologies.

As a rule, the mathematical tools, providing solution to the medical diagnostics problems, are related to the standard deterministic and statistic pattern recognition problems. The rapid development of computer-aided design technologies opened up new possibilities in this field, such as extended model integration, complex indeterministic information processing and structuralization.

The important area of research in medical diagnostics is integration of logical and precedent-related models of representation and derivation. This makes the extended functionality of computer diagnostics systems such as results validation and explanation possible. In many decision scenarios, it is important to provide the object comparison mechanism within the framework of different models. This can be achieved by eliminating some types of indeterminacy. Additionally, the transition from logical to precedent-related representation requires the construction of a large multidimensional binary objects' collection. Therefore, the information structuring is an important component in such systems, demanding particular features of the computer technologies applied. The paper proposes the solution to the decision making in medical diagnostics, while taking into consideration previously mentioned model features and information properties.

2 Medical Diagnostics as a Pattern Recognition Problem

Let us formulate a pattern recognition problem for the medical diagnostics.

Let $S_1, \ldots, S_n$ denote the set of features (symptoms). The features are used to specify patients, diseases, etc. in medical diagnostics problem. Thus, every feature must belong to a finite set of values provided by the problem context, i.e. the examination of patients.

© Springer International Publishing AG 2017
V.V. Krasnoproshin and S.V. Ablameyko (Eds.): PRIP 2016, CCIS 673, pp. 150–159, 2017.
DOI: 10.1007/978-3-319-54220-1_16

As a result, every feature S_i $(i = 1, \ldots, n)$ corresponds to a value from a set D_i. The mapping

$$x: S_1 \times \ldots \times S_n \to D_1 \times \ldots \times D_n \tag{1}$$

represents an object in space X, which is the space of possible descriptions of states of patients.

The sets of descriptions of patients can be grouped on different bases. One of the obvious bases in medical diagnostics is diagnosis, specifying the different states of patients (healthy, ill, etc.).

Let $X_1, \ldots, X_l$ $(l \in N)$ denote the groups of diagnoses, satisfying the following conditions:

$$\bigcup_{i=1}^{l} X_i \subseteq X, \tag{2}$$

$$\forall i, j \in N \ (i \neq j \Leftrightarrow X_i \cap X_j = \emptyset), \tag{3}$$

The condition (2) means that the state of every patient belongs to the space of diagnoses. Obviously, several diagnoses can correspond to one patient.

Thus, the construction of space X is specified and a binary mapping on the set $X \times L$ is given, where $L = \{1, \ldots, l\}$. For each element (x, i) of the mapping on $X \times L$, it holds that $x \in X_i, i \in L$.

In practice, the decision-making is characterized by the level of information uncertainty. In the medical diagnostics, it causes the incomplete mapping specification $X \times L$ and therefore results in errors for some $x \in X$. The information uncertainty leads to other consequences, too, but the incompleteness of information is the most important factor. The following components can be undefined:

- for some $x \in X$ the second component isn't defined;
- for some $i \in L$ the first component isn't defined.

The latter case means that the description of partition of X into subsets $X_1, \ldots, X_l$ is redundant. One can eliminate it by means of removing the classes $i \in L$, for which $X_i = \emptyset$. As a result, the problem is reduced to the initial one.

In the first case, let us suppose that there exists $X^0 = \emptyset$, for which to each $x \in X^0$ there is the corresponding second component $i \in L$. It is clear that X^0 satisfies the condition $X^0 \subset X$ as well.

Now let us specify the construction of X^0. On the one hand, knowledge can be represented by the theoretical data (e.g. from a manual) and, on the other, can be deduced from the practice. In the first case, it contains rules or regularities, which control objects $x \in X$ represented as (1). In the latter case, it contains the descriptions of the same objects (1), for which the diagnosis has been verified and, as a result, a label $i \in L$ corresponds to each such an object. These options of the set X^0 description correspond to generally accepted ways of set descriptions, i.e., they are based on the convolution principle and direct specification of the set elements.

Let X^{01} and X^{02} denote the parts of X^0, constructed as mentioned above. Similarly to the medical practice, both ways of constructing samples X^{01} and X^{02} exist independently. Let us suppose that $X^{01} \neq \emptyset$ and $X^{02} \neq \emptyset$, and the following condition holds:

$$X^{01} \cup X^{02} = X^0, X^{01} \cap X^{02} = \emptyset \tag{4}$$

Let $I_0(X^{01}, X^{02}, X^0, X_1, \ldots, X_l)$ denote all information being considered in the problem context, using conditions (1)–(4), the methodology of building the subsets, etc. It is necessary to specify an algorithm (a rule or a method) of the following type in order to solve the decision problem:

$$\forall x \in X \ (A{:}x \times I_0(X^{01}, X^{02}, X^0, X_1, \ldots, X_l) \rightarrow result) \tag{5}$$

It should be possible to interpret the result in (5) in terms of classification $X_1, \ldots, X_l$.

The above problem is called as the medical diagnostics problem with a combined training set.

3 The Basic Technological Principles of the Solution

Let us formulate the basic technological principles for solving the medical diagnostics problems with a combined training set. Initially, the most general scheme of system design, information processing and components interaction is shown (see Fig. 1).

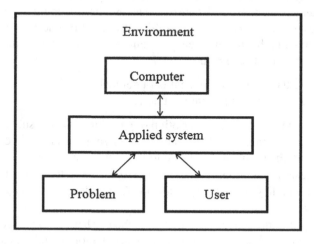

Fig. 1. The scheme of objects' interaction in the process of problem solution

This scheme does not depend on the problem domain and specifies the system components and the type of their interaction. The environment-specific limitations are common for any such system. The problem-specific features of the system are based on the interaction between the user and the information. The purpose of the applied system is to automate this interaction.

Therefore, in order to solve the technology development problem the detailed specification of the following level is provided (see Fig. 2).

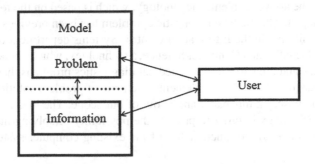

Fig. 2. The scheme of interaction between the user and information in the process of problem solution

Let's consider an arbitrary medical diagnostics problem with a combined training set, for which the type and properties of subsets X^{01} and X^{02}, the training and recognition processes requirements, the conditions for algorithms and the types of results are determined and the limitations to the cost and the ways of obtaining them are specified. Besides, the range and roles of users are defined. One needs to develop a technology for building computer systems that can provide the solution of the problem with the above limitations. The technology should:

- be based on the unified algorithmic kernel that provides possibility to use it in order to solve practical problems from different domains;
- contain means of formalized knowledge input and adjustment to a problem domain. They should include the possibilities of classes description with using different models, feature space description, training the system, etc., and provide the possibility to modify the knowledge base on the problem domain easily;
- provide the modularity of the system construction in order to guarantee the system flexibility, easy extension of its functionality, etc.;
- support the possibility to eliminate some types of indeterminacy that appear during the transition from logical to precedent-related representation of information and vice versa, and the structuring of prior information, directed at optimization of the solving problem process.

The solution of the above problem is provided at the theoretical, technological and applied levels.

- The theoretical level is specified by the effective implementation of the developed models and algorithms.
- The technological level includes problems, related to the development of computer systems for solving medical diagnostics problems with a combined training set. The

whole lifecycle of the information processing for the problems considered is provided.

- The applied level consists of using the developed library of components configured to the problem domain for solving the practical medical diagnostics problem.

For solving the above problem, a technology, which is based on the following idea, has been developed. The medical diagnostics problem has been previously considered in different definitions. In the Information Control Systems department of the Faculty of Applied Mathematics and Computer Science, a technology, which contains all necessary components and aims to solve the medical diagnostics problems, has been developed and successfully used. The implemented scheme of problem solution is different from the one proposed in this paper and is out of our scope. The idea is to extend the implemented technology in order to provide the possibility of solving the diagnostics problem and thus extend the functionality of the existing computer systems to a new level.

Several modules such as decision-making support modules, dialog organization and interface module have enhanced the technology. Below goes the brief introduction to the modules.

The following components have been added to the decision-making support module:

- Modules for transition from logical to precedent-related representation of information and vice versa.
- Object normalizing module.
- Algorithm training module.
- Module for optimizing the obtained solution based on branch-and-bound method.

The following components have been added to the dialog organization and interface building module:

- Coding module for scaled and continuous features.
- Explanation module.
- Interface module.

From the architectural point of view, the structure of the technological basis corresponds to the one that has been developed before [1–3]. The following steps outline its main modules operation.

The representation of a decision-making algorithm:

Stage 1.   The analysis of information.

> Step 1.1. The training set, which is described using logical model, is represented using precedent-related model.
> Step 1.2. The description of classes (diagnoses) is also represented using precedent-related model.
> Step 1.3. Normalization of objects.

Stage 2.   The examination.

> Step 2.1. The input data, including the description of objects and classes are formed (see Stage 1).

Step 2.2. The results of diagnostics are calculated and interpreted.

Stage 3.   Stop.

The scheme of practical system usage:

Step 1.   The information about the patient is entered (personal data and the results of the preliminary examination).

Step 2.   The localization of the disease is determined.

Step 3.   The diseased area is examined and additional symptoms are found.

Step 4.   The diagnosis is determined.

Step 5.   The results of the examination are represented in the user-friendly form.

Step 6.   The treatment scheme is formed.

Step 7.   Stop.

To verify the modules of training and optimization of algorithms, two mechanisms are used. The first one consists of executing the standard tests on the verified information. The second one is common in decision support systems and is based on the following idea: the inverse implication algorithms are implemented in the system and used to prove the monotony of direct deduction algorithms with the verified information. In order to implement these algorithms, a derivation that can answer the following questions is used in the explanation system:

(1)   How many diagnoses correspond to the current examination of the patient?

(2)   What is the maximal value of the diagnosis confirmation degree that can be obtained during the following examination of the patient?

(3)   What are the symptoms that correspond to the diagnosis?

(4)   What are the diagnoses that correspond to the chosen symptoms?

(5)   What is the list of features, for which the value of the diagnosis confirmation degree can be equal to 1?

The questions 1, 3, 4, 5 in this list can be easily implemented. The tables that show the correspondence between features and diagnoses are used for this purpose. For the question 2, an additional implementation is needed. Two inverse implication algorithms are used to answer this question and are shown below.

The algorithm for binary features:

Stage 1.   Obtaining the input data: the patient vector x, the diagnosis number i and the vector x^i from the training set, for which the current estimate for the diagnosis i is calculated.

Stage 2.   Performing the procedure of building a new vector x': those features of vector x that have already been examined with the value 1, are extended by the remaining part of vector x^i.

Stage 3.   The calculation of the upper estimate. Execute the following procedure:

Step 3.1. Perform the elimination by vector x'. As a result, a group containing diagnosis i is determined.

Step 3.2. Calculate estimates for all diagnoses from the obtained group. As a result, the upper estimate S' for diagnosis i is obtained.

Step 3.3. If this estimate is maximal in the group, output it. Otherwise, go to step 3.4.

Step 3.4. Replace all values 0 in vector x' by indefinite ones and calculate the lower estimate S for the obtained vector. Output estimate S.

Stage 4.   Stop.

For all cases output the patients, for which this estimate is obtainable.

The algorithm for scaled features:

Stage 1.   Obtaining the input data: the patient vector x and the diagnosis number i for the group of diagnoses containing i.

Stage 2.   Calculating the lower S and upper S' estimates by a vector, which allows obtaining the maximal estimate g in the diagnosis i:

Step 2.1. Replace every feature, as soon as its value is 0 or indefinite, by 1. If the estimate g rises, perform the correspondent replace in the patient vector. As a result, a new patient vector is obtained.

Step 2.2. Calculate estimate S' for the given patient vector by class i.

Step 2.3. Calculate estimate S by class i and the patient vector that differs from the built one only by positions, which have value 0: one need to set an indefinite value in these positions.

Stage 3.   Identifying the possibility to show the upper estimate S':

Step 3.1. If $S' = 1$, then output S'.

Step 3.2. If $S' < 1$, then extend the initial (current) vector by those values 1, that appeared during stage 2, and perform a new division into groups.

Step 3.3. Calculating estimates S and S' by all diagnoses. Afterwards, analyze the estimates to determine, which of the two estimates can be shown.

Stage 4.   Stop.   •

As in the algorithm for binary features, for all cases output the patients, for which this estimate is obtainable.

In the process of solving the medical diagnostics problem the dialog organization and interface building module, the decision-making support module and some third-party modules like browser or Web-server are used. The sequence of their work, based on the developed technology, can be described as follows.

The scheme of using the technology for solving medical diagnostics problems:

1. The dialog organization and interface building module provides the dialog with the user.
2. The decision-making support module solves the problem using the knowledge file and the algorithmic kernel.
3. The dialog organization module outputs the results and their estimates.
4. Stop.

This scheme is shown in Fig. 3.

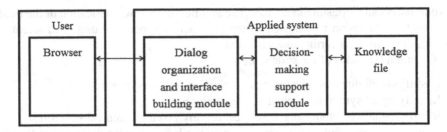

Fig. 3. The scheme of using the technology for the pattern recognition problem solution in medical diagnostics

Initially, the dialog organization and interface building module converts the requirements of the feature space to the questions addressed to the user and converts the answers to the feature values. The user can inquire the system about the additional information. This module is also used to output the solution as a list of found diagnoses with their estimates and allows using the means of explanation of the obtained results.

The decision-making support module represents the central part of the system and implements the algorithmic kernel. In the process of solving the problem this module controls the examination and calculates the estimates. The necessary information is obtained from the knowledge file, which is created during the development. In this file, features and classes are formally listed, the description of classes is given as a set of logical rules and training sets, additional options of algorithms are specified.

The standard methodology of using the algorithmic kernel consists of the following steps: first the algorithm is chosen and then its configuration and verification are performed. If the results are erroneous, the algorithms are corrected. The number of objects in the prior and test samples is usually limited by the size of information that is available to the researchers. Further, during the commercial operation stage, the base algorithms can be corrected in the case of receiving new information.

4   The General Description of System EXTRA

The system EXTRA has been developed based on the above technology for solving problems of diagnostics orthopedics diseases. In this chapter, the system is described similarly to [2–6]. The main purpose of the system is support of decision-making in the areas of orthopedics, sporting traumatology and rehabilitation, providing the doctor with the continuous informational help during diagnostics and choosing the treatment tactics based on general and special medical knowledge.

The base information in the system is the diagnosis with fixed one or several (alternative) treatment schemes. The final rehabilitation protocol is made depending on the current state of patient. The information is of a recommended character and can be corrected at user's discretion.

The architecture of the system is based on the client-server technology and distributed data processing. The client part includes end-user-oriented functions. The server components of the system provide database services and knowledge bases controlling.

Several users can work simultaneously. The architecture allows unification of the treatment and rehabilitation methods exchange of experience between users and scientific data mining with their further interpretation.

The main functions of the system are:

- Identifying all diagnoses including any given symptom/symptoms;
- Identifying all symptoms typical for this diagnosis;
- Identifying all symptoms that are typical for any given anatomical structure;
- Searching the description of the treatment methods based on the diagnosis and symptoms for a specific patient.

Figure 4 shows the general structure of system EXTRA.

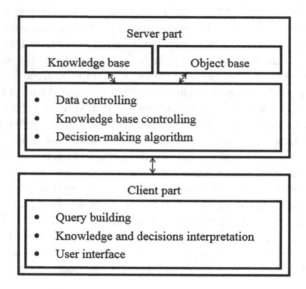

Fig. 4. The structure of system EXTRA

The process of solving, i.e. finding the diagnosis, is based on the following ideas.

The problem domain is divided into independent subsystems – localizations, where the decision support is made independently. There exist classical examination methods that allow obtaining the set of features, which is enough for the diagnostics. They are divided into subsets that in total describe some diagnosis. The localization is defined by the set of diagnoses. Each diagnosis is connected with a set of rules that determine some prior information for the diagnosis. To form the rules, logical operations are used. Each diagnosis corresponds to a set of possible actions for treatment.

The knowledge base includes descriptions of features, diagnoses, formulas, etc. The symptoms, which describe the state of patient, make a system of binary features. The features are ordered as a treelike structure, which corresponds to the order of the patient examination. The rules of calculating the diagnoses are stored as logical formulas using logical operations AND, OR and XOR in the feature space.

In the system EXTRA, the methods, which are based on conversion of the problem to a standard pattern recognition problem, are used. For this purpose, an algebraic structure of information is introduced. Localizations correspond to a list of diagnoses, a list of features and a map of examination, which represents a table that shows the usage of features in the descriptions of diagnoses. The part of training set, which is represented using the logical model, is converted to the precedent-related model. The resulting information is joined with the part that has initially been represented using the precedent-related model. Then a special pattern recognition algorithm is used on the joint information. The training set for this algorithm is built using the operations of the logic algebra. Thus, the solving of a problem with combined prior information is reduced to the solving of a pattern recognition problem that uses only the precedent-related model. The extension of the functionality of the system EXTRA has allowed improving the quality of diagnostics process while searching the diagnosis and choosing the treatment tactics in the area of sports traumatology.

5 Conclusion

The paper outlines a construction of algebraic base for the modeling and implementation of applied medical diagnostics support systems.

In addition, the component and functional structure of an artificial intelligence computer system EXTRA for diagnostics diseases is presented in the paper. The system implements the decisional kernel based on the algorithms for solving the pattern recognition problems with a combined training set.

References

1. Ablameyko, S.V., Krasnoproshin, V.V., Obraztsov, V.A.: Models and technologies for pattern recognition with application in data mining. Vestnik BSU. Ser. 1 Ph. Math. C. Sc. **3**, 62–72 (2011) (in Russian)
2. Krasnoproshin, V.V., Gafurov, S.V.: A program technology of building systems for solving pattern recognition problems with complicated structure. Artif. Intell. **1**, 30–37 (2008). (in Russian)
3. Krasnoproshin, V.V.: Theoretical Basis and Technologies for Constructing Recognition Systems by Precedence. Dr. Sc. Thesis 05.13.17. Minsk (2007) (in Russian)
4. Ablameyko, S.V., Krasnoproshin, V.V., Obraztsov, V.A.: The experience and main results of information control systems department (Belorussian State University) in the field of theory and practice of pattern recognition. In: Proceedings of Crimea International Mathematics Conference (CIMC-2013), vol. 3. Simferopol (2013) (in Russian)
5. Ablameyko, S.V., Krasnoproshin, V.V., Obraztsov, V.A.: Pattern recognition and image analysis: theory and experience of solution of practical problems. In: International Congress of Computer Science: Information systems and Technologies, pp. 434–444. BSU, Minsk (2013) (in Russian)
6. Krasnoproshin, V., Lositskiy, Ye., Obraztsov, V., Vissia, H., Gutnikov, S., Popok, S.: Intelligence system of decision support in sporting traumatology. Bulletin of NTU "KhPI". Comput. Sci. Modeling **31**, 106–111 (2010) (in Russian)

The Algorithm for Monitoring of Evolution of Stem Cells on Base Integral Optical Flow

O. Nedzvedz[1] and S. Ablameyko[2(✉)]

[1] Belarusian State Medical University, Minsk, Belarus
olga_nedzved@tut.by
[2] Belarusian State University, Minsk, Belarus
ablameyko@bsu.by

Abstract. In this paper, we propose a novel algorithm for monitoring of steam cells by using integral optical flow. The algorithm is based on construction of field of integral optical flow and its structure allows one to describe stages of cell's evolution. The algorithm allows to analyze a motion send the internal structure of the cell and cell's conglomeration. Definition of motion of such object by integral optical flow allows one to make a detailed prognosis of evolution of such objects.

Keywords: Optical flow · Image monitoring · Dynamic object · Cell's evolution

1 Introduction

One of the important tasks in situation monitoring in video sequences is tracking of cells. In real life a cell are dynamical objects [1] than can changed alone and aggregation sets [2]. Such an object has an internal structure consisting of interacting dynamic elements. Dynamic objects combine movement of objects as a whole with the movement of internal structures, their aggregation or division.

Methods for detection and tracking of dynamic objects can be divided into tracking-based methods that examine trajectories of objects and tracking-free methods based on examination of visual features such as differences in color or brightness changes [3, 5]. Permanent transformation of dynamic images like change of size, shape and orientation makes difficult using of this method for these objects. Detection and tracking of dynamic objects is a more complicated task compared with tracking of simple moving objects. Sometimes it is difficult to determine the boundaries between inner parts of dynamic object due to their occlusion and merging, for example, if a group of people moves in the same direction. Methods like thresholding techniques [6, 7], neural networks and probabilistic object models [8, 9] are used for these objects.

Methods for tracking of dynamic objects use geometrical information about motion of objects over time. These methods do not depend on an object type and shape. Usually they are based on dynamic active contour models [10].

One of the most promising methods for motion analysis of dynamic objects based on using an optical flow. Optical flow belongs to region-based methods [11]. This method

© Springer International Publishing AG 2017
V.V. Krasnoproshin and S.V. Ablameyko (Eds.): PRIP 2016, CCIS 673, pp. 160–170, 2017.
DOI: 10.1007/978-3-319-54220-1_17

allows to get the distribution of velocities and directions of points of object from shift of these points between two images. Optical flow is widely used for investigation of different types of motion like translation of moving object relative to the static or dynamic background and another moving objects; rotation of the object relative to the axis. However, case of dynamic objects with unstable shape or internal structure noise and random changes has a great influence on the optical flow in video sequence. As a result, the structure of motion vectors can be unclear.

In this paper, we propose a novel algorithm for monitoring of evolution of cells as dynamic objects by using integral optical flow. The algorithm is based on construction of field of integral optical flow. Structures in this field allow one to describe stages of motion for cells.

2   Scheme of Image Processing for Dynamic Objects

The basic factor for description of state of complex dynamic objects is the shape changing. Such objects change their shape incessantly. Therefore, task of monitoring of dynamic scene evolution can be solved by controlling the shape characteristics.

The general scheme of such monitoring consists of five stages (Fig. 1):

- image acquisition and preprocessing;
- scene image segmentation and simple objects detection;
- measurement of characteristics;
- determination of laws for common description;
- object classification.

The first stage is the image acquisition and preprocessing for improvement of the image quality. As a rule, the image acquisition process depends on the speed requirements. This property can also be used for image improvement. It is possible to use frames averaging through very short time for image quality improvement. In the result, procedure of acquisition is irregular and has differentiation in time. The quality image is constructed by averaging frames that capturing in short range of time. In this way, a pixel on the image at every iteration of capture is corrected as:

$$I''(x,y) = \frac{I'(x,y) \cdot (n-1) + I(x,y)}{n},$$
(1)

where (x,y) is a coordinate of processed pixel, $I''(x,y)$ is a new value of intensity of pixel, $I'(x,y)$ is a previous value of pixel intensity, $I(x,y)$ is a value of pixel intensity in current captured frame, n is count of capture iterations in a short time.

This acquisition process is stopped by checking of distortion error that is calculated for every new frame as follows:

$$\sigma''(x,y) = \frac{\sigma'(x,y) \cdot (n-1) + I''(x,y) - I(x,y)}{n} < \varepsilon,$$
(2)

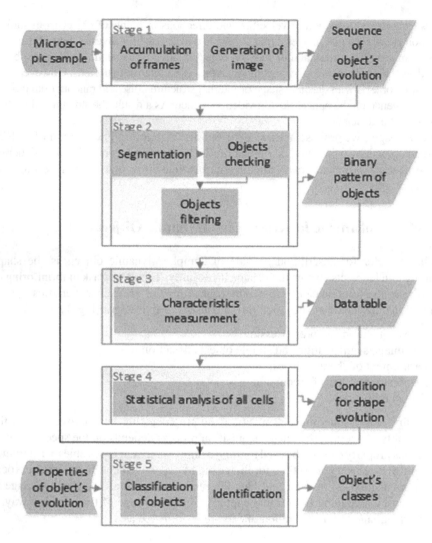

Fig. 1. General scheme for image processing with dynamic objects in video sequence

where (x,y) is a coordinate of processed pixel, $\sigma''(x, y)$ is a new error of intensity of pixel, $\sigma'(x, y)$ is the error of pixel intensity for previous iteration, $I''(x, y)$ is a new value of intensity of pixel, $I(x,y)$ is a value of pixel intensity in current captured frame, n is a count of capture iterations in short time, ε is an accuracy for error detection.

Usually the sequence of such images is viewed as movie with speed about 30 fps (Fig. 2). Acquisition procedure is performed at certain time intervals. Thereby each frame of a video sequence was generated by procedure of averaging for images that have been captured in a very short time interval after a certain waiting period.

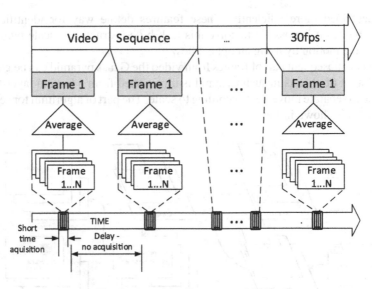

Fig. 2. Process of image acquisition

3   Integral Optical Flow in Dynamic Scene

The concept of optical flow allows one to study the motion in a scene. Usually in cytology dynamic scene consists of various moving cells.

The optical flow allows one to estimate motion between two frames of video sequence. It is based on determination of displacement for intensity I around every pixel of image that is changing in range of time from t to $t + \Delta t$. This difference is defined in near neighborhood. Therefore optical flow estimates displacement (dx,dy) for every pixel in space between two frames of video, where:

$$I(x, y, t) = I(x + dx, y + dy, t + dt); \tag{3}$$

In case of small displacement between two frames this value can be expressed as:

$$I(x + dx, y + dy, t + dt) = I(x, y, t) + \frac{\partial I}{\partial x}dx + \frac{\partial I}{\partial y}dy + \frac{\partial I}{\partial t}dt \tag{4}$$

From this equation, the equation of the optical flow can be obtained:

$$\frac{\partial I}{\partial x}V_x + \frac{\partial I}{\partial y}V_y + \frac{\partial I}{\partial t} = 0 \tag{5}$$

where V_x is horizontal component of velocity, V_y is vertical component of velocity, $\frac{\partial I}{\partial x}, \frac{\partial I}{\partial y}, \frac{\partial I}{\partial t}$ are partial derivatives.

In many cases, visual movement is displacement of objects located at different depths of the image. Therefore, after the separation of objects in layers their layered motion can be described more efficiently. These features define way for identification of different objects. It is possible to solve this task by constructing a scale pyramid and estimating the motion by using an optical flow.

For every subsequent pair of frames from video the Gauss pyramid can be generated. The resolution of objects in such pyramid is diminished from a bottom layer to top. It allows one to separate movement according to scale. The part of algorithm for separation of movement is shown in Fig. 3.

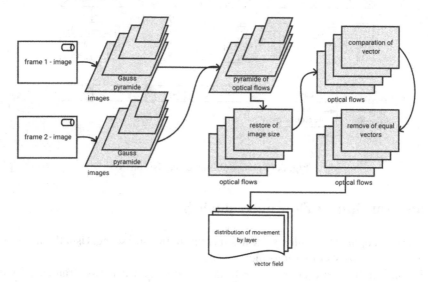

Fig. 3. Algorithm of dynamic object classification by pyramid of integral optical flow

In this way, levels of scales pyramid of optical flow have different resolution for vectors. For lower layers, vectors accumulate all values and directions of vectors started from upper layers. Small movements are blurred on downsize layer. The basic movement of large objects is represented in all layers of pyramid. The scaling of layers of pyramid does not allow to make analysis of vectors values. It is necessary to change size of every layer of pyramid to size of button layer for comparison of vectors of optical flow. After that, the analysis of vector fields at neighboring layers of pyramid is made to define vectors with the same characteristics.

In the result of such procedure, every layer includes information about only one class of movement. This part of algorithm allows one to classify motion of objects in video. However, sometimes quality of result of such algorithm is not high because background usually includes many dynamic objects. For such objects, the value of vector of optical flow can have the same value as for real objects and classification of motion includes errors.

The best solution of this problem is accumulation vectors of optical flow to integral optical flow. Integral optical flow for each pixel is formed as a result of integration of values of the optical flow over the whole video sequence:

$$\vec{I}(x, y) = \sum_t \vec{V}(x, y) \tag{6}$$

where I is the integral optical flow, V is the vector of optical flow, (x,y) is a coordinate of point in the image space.

Different categories of objects have various characteristics of integral optical flow. Objects of first category are elements of background. They make movement around their point of location. As a result, vectors of integral flow have small values. Realization of this part of the algorithm is based on accumulation of integral optical flow during sequence processing frame by frame (Fig. 4).

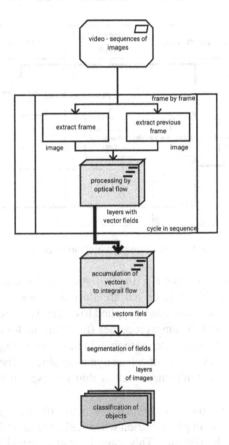

Fig. 4. Application of integral optical flow for motion analysis

Vector field of integral optical flow is accumulated for every resized layer from the pyramid. In the result, background movements are removed and basic motion is strengthened. The operation of accumulation of integral optical flow can include operation for comparison of vectors between layers. Additional control allows one to decrease errors for classification of dynamic objects. After that the segmentation is started for object classification.

4 Evolution of Cells

Complex changes in the form and content of a dynamic object occur in time. We determine these changes as an evolution of an cells.

The analysis of evolution of cells includes five stages: aggregation, simple motion, intermediate stage, mitosis (division) and apoptosis (destroying) (Fig. 5). In this scheme, dot lines show evolution of cell as one object. Solid lines demonstrate generation state "one objects" object from other ones.

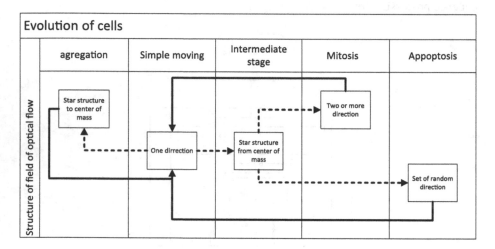

Fig. 5. Scheme of evolution of cells

The basic component of motion is a simple motion. In this case, a cell is represented as a common one body. All parts of cell are shifted in one direction with common velocity. Usually such description can be applied to cell as a more complex object where different parts with different dynamic properties (for example for nuclear or organelles) can be united into a common geometrical region. In this case, a simple motion is a central element in the scheme of evolution of cell as dynamic object. The integral optical flow of such object has vectors with one direction that correspond to direction of object motion.

The object evolution from this state has two ways of aggregation or intermediate state. The aggregation is a simple operation where cells are united in conglomerate and has common and fussy border into. This stage is characterized by a star structure of vectors of integral optical flow that has direction to center of uniting. Intermediate stage is very unstable. Usually it is changed within short time. Vectors of integral optical flow are also characterized by a star structure. In this case, they have direction from a center of mass into a cell. In this case, cell tries to fill a maximum region. After this stage, cell going to apoptosis or mitosis. In mitosis the cell is divided to two or more cells, in this stage an integral optical flow is characterized by motion of new objects and it usually has a few basic directions from center of mass of source cell. Every basic vector of this flow characterizes simple motion of a new object. Then it is possible to spend up analysis of dynamic properties for every new object separately. In case of apoptosis, the cell is divided to many small objects. The field of integral vector flow has too many different directions and small absolute values of vectors. In this case, it is possible to spend analysis of every new object but usually it does not make sense.

5   Results and Discussion

For live stem cells, there are few types of motion. We extract four basic types: simple motion, intermediate stage, apoptosis, mitosis and aggregation of objects to one conglomerate.

5.1   Cytological Scene Analysis

We understand a simple motion as a geometrical shifting an object without changing its shape. In this case, if the object is shifted in only one direction, the vector of optical flow increases for every points of this object. In the same time, this value is decreasing for a stochastic motion. Therefore, it is possible to define regions of useful movement on base of module of vector of integral optical flow.

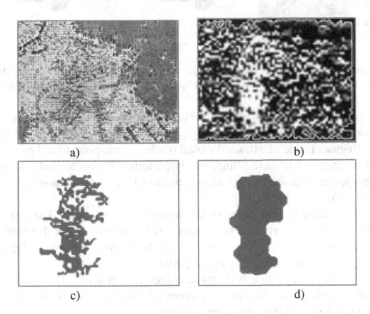

Fig. 6. Using of integral optical flow for definition of region of interest for video with moving car: (a) last image from source video with vectors of integral optical flow; (b) map of modules of integral optical flow; (c) regions of active motion from maximums of integral optical flow; (d) region of object's motion

Vectors of integral optical flow allow one to extract a basic motion in a scene. In Fig. 6b, image is constructed from absolute values of integral vector flow. This image can be binarised by thresholding (Fig. 6c). Correction by morphological filtration (close and open) removes noise in region of motion (Fig. 6d).

5.2 Cell Image Analysis

For determination of region of cells in a microscopy sample, image segmentation is usually used. This is very important procedure, because a system produces a solution about object class on a base of extracted regions. It should be noted that the integrated optical flow allows one to eliminate dynamic objects that belong to a background.

Segmentation started from a bottom layer. This process consists of two parts. The first part is an analysis of absolute vector value. In this case, map of such characteristics looks like image that can be analyzed by image processing methods (Fig. 7).

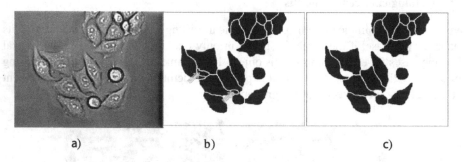

a)   b)   c)

Fig. 7. Segmentation of button layer of image with live cells: (a) source image, (b) simple segmented image, where arrows show to false objects, (c) binary image after object's filtering

An operation of morphological filtration (close and open) and small object removing allow to correct these borders. The thinning operation allows one to get a closed border of objects. Regions inside of extracted closed borders correspond to cell patterns on an image. After operation of hole filling, binary patterns are constructed on the image. Unfortunately, these binary patterns include regions that do not correspond to the real objects (Fig. 7b).

Such regions have similar intensity levels over the whole area. The dispersion or bandwidth of intensity levels in every region can be checked. After that regions with large values of these characteristics (Fig. 7c) can be removed by object filtering. As a result, binary image consists of only object patterns.

The fifth stage of algorithm is dynamic object classification. There are five basic types: normal moving (Fig. 8), critical or intermediate stage (Fig. 9), apoptosis of cell (Fig. 10a), mitosis of cell (Fig. 10b), aggregation.

Fig. 8. Normal stage of cell and integral optical flow

Fig. 9. Intermediate stage of cell and integral optical flow

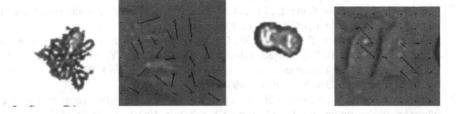

Fig. 10. Apoptosis (a) and mitosis of cell and integral optical flow (b)

Processes of mitosis (Fig. 10b) or apoptosis (Fig. 10a) are going from intermediate stage. Vectors of optical flow have structure like a star and is going from center of any event. The same situation as in outdoor scene there is in evolution of living cells. They have intermediate stage of evolution when cell (Fig. 9) is growing and vectors of optical flow have a star structure and after that cell can be destroyed (stage of apoptosis) or divided into two new cells (stage of mitosis). In the first situation, vectors of optical flow have various random directions (Fig. 10a). In the second situation, they have two opposite directions (Fig. 10b).

In the result, it is possible to describe all stages of cell evolution by field of integral optical flow.

6 Conclusion

Monitoring of the evolution of live steam cells is a complex procedure. For its implementation it is necessary to take into account its shape at various stages. Changing of such objects is performed by acquiring a rounded shape. After this change, the object is divided or destroyed. Based on these features, an efficient algorithm for automation of object monitoring and description was developed. Using this algorithm, the quality of solution of many tasks for monitoring complex dynamic objects can be improved. Definition of stages of cells development by integral optical flow allows one to make a detailed prognosis of evolution of such objects and can be applied for various automatic monitoring systems for complex motion investigation.

Changing of structures in integral optical flow allows one to use monitoring of motion for various tasks. The proposed method was designed for investigation of new possibilities in motion and improve the quality control of cultivated tissue.

Acknowledgments. This work is supported by the Program BRFFR F16R-180.

References

1. Wang, L., Hu, W., Tan, T.: Recent developments in human motion analysis. Pattern Recogn. **36**(3), 585–601 (2003)
2. Huh, S., Eom, S., et al.: Mitosis detection of hematopoietic stem cell populations in time-lapse phase-contrast microscopy images. In: Proceedings of the 9th IEEE International Symposium on Biomedical Imaging, pp. 390–393 (2012)
3. Qiu, C., Zhang, Z., Huanzhang, L., Luo, H.: A survey of motion-based multitarget tracking methods. Prog. Electromagnet. Res. B **62**, 195–223 (2015)
4. Yuan, G., Zhang, J., Han, Y., Zhou, H., Xu, D.: A multiple objects tracking method based on a combination of camshift and object trajectory tracking. In: Tan, Y., Shi, Y., Buarque, F., Gelbukh, A., Das, S., Engelbrecht, A. (eds.) ICSI 2015. LNCS, vol. 9142, pp. 155–163. Springer, Heidelberg (2015). doi:10.1007/978-3-319-20469-7_18
5. An, M.-S., Kang, D.-S.: A method of robust pedestrian tracking in video sequences based on interest point description. Int. J. Multimedia Ubiquit. Eng. **10**(10), 35–46 (2015)
6. ELHarrouss, O., Moujahid, D., Elkaitouni, S.E., Tairi, H.: Moving objects detection based on thresholding operations for video surveillance systems. In: IEEE/ACS 12th International Conference of Computer Systems and Applications, Marrakech, pp. 1–5 (2015)
7. Vijayakumar, P., Senthilkumar, A.V.: Threshold based filtering technique for efficient moving object detection and tracking in video surveillance. Int. J. Res. Eng. Technol. **5**(2), 303–310 (2016)
8. Watter, M., Springenberg, J., Boedecker, J., Riedmiller, M.: Embed to control: a locally linear latent dynamics model for control from raw images. In: Advances in Neural Information Processing Systems, pp. 2728–2736 (2015)
9. Sutskever, I., Hinton, G.: Learning multilevel distributed representations for high-dimensional sequences. In: International Conference on Artificial Intelligence and Statistics, pp. 548–555 (2007)
10. Lecca, M., Messelodi, S., Serapioni, R.: A new region-based active contour model for object segmentation. J. Math. Imaging Vis. **53**(2), 233–249 (2015)
11. Yang, J., Li, H.: Dense, accurate optical flow estimation with piecewise parametric model. In: IEEE Conference on Computer Vision and Pattern Recognition, pp. 1019–1027 (2015)

Hyperspectral Data Compression Framework for Earth Remote Sensing Objectives

Alexander Doudkin[1]([✉]), Leonid Podenok[1], and Dmitry Pertsau[2]

[1] United Institute of Informatics Problems, Minsk, Republic of Belarus
doudkin@newman.bas-net.by, podenok@lsi.bas-net.by
[2] Belarusian State University of Informatics and Radioelectronics, Minsk, Republic of Belarus
DmitryPertsev@gmail.com

Abstract. The hyperspectral data compression framework to well investigate various compression models is presented. Results received with arithmetic encoder, context-adaptive QM-encoder, adaptive Huffman encoder are adduced. As a test data the Maine frame set from the AVIRIS freely available data was used. The received results testify the efficiency of the proposed framework in comparison with some alternative lossless compression algorithms.

Keywords: Hyperspectral data · Fourier Transform Imaging Spectrometer · Arithmetic coding · Context-adaptive QM-encoder · Adaptive Huffman encoder · AVIRIS

1 Introduction

Remote sensing is a method of acquisition of information about objects or areas without making physical contact with them. To acquire information, the specialized film-equipment is mounted on board of the satellite or the plane, flying over the target area. The equipment fixes reflected radiation from Earth surface in various spectral bands, and then converts it into digital form for transmission obtained data to the processing station.

Operating range of wavelengths of the remote sensing process is determined by specific problem facing the mission. There are systems dealing with radiation from micrometers (visible optical) to meters (radio wave). Multispectral systems usually operate within several non-overlapped spectral bands while hyperspectral ones treat with hundreds adjacent narrow bands.

Depending on the type of orbital sensor there are distinguished multispectral (e.g., Landsat, IKONOS, Rapid Eye, etc.) and hyperspectral (e.g., AVIRIS) ones. The main difference between them are the number of bands and their location order over spectrum. A multispectral sensors cover the spectrum from the visible up to the longwave infrared. They do not provide the continuous spectral range of an object but some discrete regions. Unlike these hyperspectral sensors deal with narrow spectral bands over a continuous spectral range, and produce the real spectrum of all pixels in the scene. So a sensor with 20 bands can be both hyperspectral when it densely covers whole range and multispectral when its bands did not adjoin.

V.V. Krasnoproshin and S.V. Ablameyko (Eds.): PRIP 2016, CCIS 673, pp. 171–179, 2017.
DOI: 10.1007/978-3-319-54220-1_18

2   Trends in Remote Sensing System Progress

Trends in the evolution of remote sensing systems show that the emphasis is shifted to hyperspectral direction. However, the wide practical use of aerospace monitoring on the one hand hampered by the lack of sufficient number of satellites and aircraft equipped with appropriate spectrometers, and on the other hand there are difficulties associated with transmission, processing and interpretation of large amount of data generated by these devices.

Traditional data representation (classic hypercube) transmitted to the processing center are three-dimensional cube (Fig. 1) with the following resolution characteristics:

- spatial which determines the underlying surface details;
- spectral which determines width of spectral band;
- radiometric which determines number of signal levels that the sensor can register.

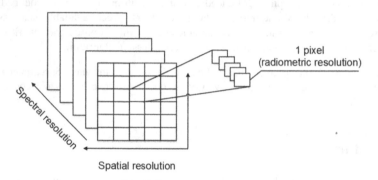

Fig. 1. Remote sensing data structure

The structure leads to need to transmit a huge data volume from orbit to the Earth and the problem of data compression becomes very acute and comes to the edge of technology. For example, volume of the AVIRIS [1] data sets which are used for compression algorithms and software research and development has following characteristics – 680 columns width (for AVIRIS Maine), 224 spectral samples per pixel, 12 bits per sample. This leads to more than 223 kB data per row to be transmitted over radio link. Taking into account the characteristics of modern radio links and the fact that the scanning is carried out in continuous mode the primary requirements to data compression algorithms are high compression ratio and low computation complexity. The last is concerned with limited onboard hardware capabilities.

3   Fourier Transform Image Spectrometer

There are some approaches to form the hyperspectral response for single pixel of image. Classical solution is using the prism or diffraction grating with swinging mirror scanning along row and orbital movement as column directions. Particular kind of hyperspectral equipment is the Fourier Transform Imaging Spectrometer (FTIS) based on any interferometer, for example, Sagnac one [2] (Fig. 2). Unlike equipment fulfilling direct spectral measurements FTIS outcome is interferogram that requires special processing to obtain the same spectral image cube with dimensions equal to rows × cols × spectral response. That special processing is one of Fourier transforms, e.g. cosine one. There are two approaches – to transmit raw FTIS data over radio link and fulfill Fourier transform on receiver side, and fulfill Fourier transform onboard and transmit ready spectral data. Our work with respect to FTIS deal with the first case – raw data are compressed and transmitted; all remaining work is fulfilled by receiver.

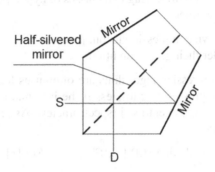

Fig. 2. Sagnac interferometer

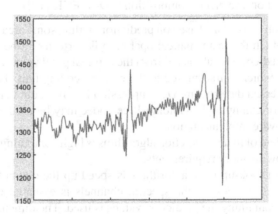

Fig. 3. Modulation function sample

To test the compression algorithms some technique of synthesis of Fourier inter-ferograms from AVIRIS hyperspectral data has been developed. Modulation function shown in Fig. 3 is formed on the basis of real data or is synthesized mathematically:

$$u_{i,j}^k = \frac{v_{i,j}^k \cdot m_k}{v_{max}^k \cdot m_{max}} \tag{1}$$

where $u_{i,j}^k$ and $v_{i,j}^k$ – value in a layer k of pixel (i, j) of the initial and modulated cubes, respectively; v_{max}^k – the maximum value in a layer k ; m_k – the maximum element in a layer k ; m_{max} – maximum element of all cube. The received value is rounded up to closest integer number.

4   Review of Compression Algorithms

After analysis of the literature following two classes of hyperspectral data compression algorithms have been identified:

- using commonly known compression techniques;
- using adaptation under their given condition.

The first class of compression algorithms are often uses lossless and near lossless methods. Algorithm compresses near lossless if the information loss does not exceed the noise induced by the equipment used (spectrometer). As part of this approach are the following basic classes:

- algorithms based on prediction (LP [3], FL [4], SLSQ [5], CCAP [6], M-CALIC [7]);
- algorithms based on table look-up (LUT [8], LAIS-QLUT [9]);
- algorithms based on wavelet decomposition (3D-SPECK [10]).

The compression algorithms based on prediction outline some area in the vicinity of the pixel within which the mathematical operation is performed (the prediction step). The result of the prediction is subtracted from the original pixel value and the prediction error is generated, which is transmitted to the entropy coding unit. The output of that unit form is compressed data stream. Decompression is carried out in reverse order. As a coding system, for example, the Golomb-Rice codec may be used or any arithmetic one allowing hardware implementation.

The main problem of many prediction algorithms is high computing complexity. But algorithms need low memory requirements.

The objective of look-up table algorithms is speed up the computations using the fact that the correlation between the spectral channels is essential enough. For this purpose, the table containing the prediction values is used. The dimension of that table is equal to number of spectral channels multiplied by the maximum value for the given radiometric resolution.

The prediction process for current `sample[row][col][layer]` is:

```
pred[row][col][plane] =
LUT[layer][data[row][col][plane-1]];
LUT[layer][data[row][col][plane-1]] =
data[row][col][plane];
```

where plane is spectral channel index. The resulting value `pred` will be considered as predicted. Further processing is equivalent to the prediction algorithm.

The algorithms based on the discrete wavelet transform is the most demanding to computing resources. This class of algorithms requires to preliminary transform every spectral plane to the spatial frequency domain. Then first are encoded the most significant (high-frequency) of the wavelet coefficients and the least significant ones are coded in the last.

This approach allows both lossless (when all the wavelet coefficients are encoded), and controlled lossy compression. The main drawback of the approach is the computational complexity associated with the data cube transformation to the frequency domain, as well as requirements to memory bandwidth because will be random memory access when processing wavelet coefficients.

The other approach is based on an essential redundancy of the generated data that is caused by high spectral resolution. The algorithms of the class are based on the following simplifications:

- requirements to necessary spectral characteristics are known. In this case it is possible to transmit only necessary spectral channels or do not transmit uninformative ones (which could occur, for example, due to bad atmospheric condition), i.e. reduce the hyperspectral case to multispectral one;
- perform a full or a partial analysis of received data and transmit the result, but not the data itself. Unfortunately, this approach is difficult to implement on satellite board. Nevertheless, the advantage of approach is that the required data transmitted to the Earth.

5   Features of Hyperspectral Data

When designing the compression algorithm some correlation characteristics of hyperspectral data was studied allowing to evaluate the similarity between pixels and channels. The following formula are used to determine the spectral (2) and the spatial (3) correlation [11]:

$$c_{u,v} = \frac{\sum\limits_{i=1}^{M}\sum\limits_{j=1}^{N}\tilde{x}_{i,j,u} \cdot \tilde{x}_{i,j,v}}{\sqrt{\sum\limits_{i=1}^{M}\sum\limits_{j=1}^{N}\tilde{x}_{i,j,u}^2 \cdot \sum\limits_{i=1}^{M}\sum\limits_{j=1}^{N}\tilde{x}_{i,j,v}^2}} \tag{2}$$

$$c_k(i,j) = \frac{C(i,j)}{\sqrt{C(i,i) \cdot C(j,j)}} \tag{3}$$

where $\tilde{x}_{i,j,k} = x_{i,j,k} - \bar{x}_k$, $x_{i,j,k}$ – value of pixel with coordinates (i,j) in the spatial slice of the channel k, $\bar{x}_k$ – a population mean in the channel k, M and N – width and height of the channel in spatial area, $C = \mathrm{cov}(X)$ – covariance matrix.

As can be seen from Figs. 4 and 5, the correlation between closely spaced interferogram tends to one. However, there are areas the correlation between which tends to zero. This is the result of weather condition, for example, part of the far infrared radiation is absorbed by water vapor and carbon dioxide. Respectively if the spectrometer covers the appropriate range of spectrum some notches may be occur on spectrogram.

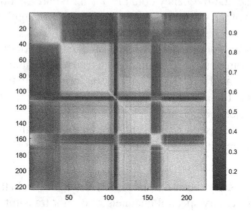

Fig. 4. Spectral correlation

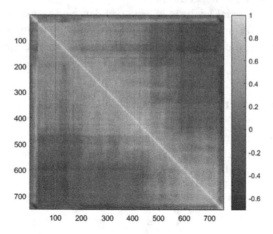

Fig. 5. Spatial correlation

The main requirements for the compression algorithm are:

- universality that means possibility of using the traditional spectral hypercube and Fourier-interferogram formed one;
- lossless compression that is caused by insufficient information available related to influence of data loss on the quality of the interferogram hypercube recovery;
- computational simplicity of the algorithm and the possibility of parallel processing.

6 Hyperspectral Data Compression Framework

In accordance with aforementioned features and requirements compression algorithm has been developed. That algorithm consists of preprocessing of each interferogram sample plane, reduction of correlation degree between the interferogram sample plane, and an encoder.

Figure 6 shows an example for three interferogram sample planes. The total number of sample planes is divided into subset of fixed size $m < M$ which are input of a compression algorithm. Recommended $m = 10 - 15$ sample planes. This recommendation is related with correlation between the sample planes (Figs. 4 and 5).

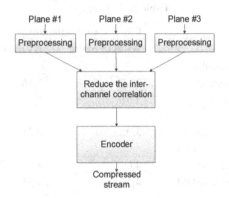

Fig. 6. Compression system model

As the preliminary step lossless wavelet decomposition is used. This allows to reduce data redundancy, and appends the ability to control the compression process (model adaptation for lossy compression case).

To reduce the degree of correlation between adjacent sample planes the difference method is used.

The final stage of the algorithm is the result encoder to compressed stream. The following options are examined as the encoder:

- adaptive Huffman encoder;
- arithmetic encoder;
- contextadaptive QM-encoder.

7 Framework Testing

AVIRIS (Hawaii, Maine) set was used to test compression model. AVIRIS is a standard in the field of hyperspectral data compression research. The sensor allows to capture images with a spatial resolution of 20×20 m per pixel in the spectral region from 400 nm to 2500 nm. Channel bandwidth is equal to 10 nm (224 spectral channels). The sensor uses 12-bit analog-to-digital converter.

To form interferogram from AVIRIS hypercube data modulation method was used (1), the modulation function shown in Fig. 3.

Test data comply with the following parameters:

- radiometric resolution – 12-bit positive integer;
- spatial image resolution – 512 lines, 680 columns for AVIRIS Maine and 614 columns for AVIRIS Hawaii;
- number of spectral channels for classic hypercube – 224;
- number of interferogram sample planes generated – 256. (Table 1)

Table 1. Test results

	Compression ratio, times	
	AVIRIS Maine	AVIRIS Hawaii
Classic hypercube		
Arithmetic encoder	2,98	3,16
Context-adaptive QM-encoder	2,91	3,11
Adaptive Huffman encoder	3,02	3,21
Fourier interferogram		
Arithmetic encoder	4,69	4,62
Context-adaptive QM-encoder	4,50	4,55
Adaptive Huffman encoder	4,78	4,70

The advantages of the proposed compression algorithm are:
- universality – it is possible to apply algorithm to the classic hypercube and interferogram one;
- possibility of parallel processing;
- computational simplicity (there are no arithmetic operations with high latency, i.e. multiplication and division).

8   Conclusion

The received results testify the efficiency of the proposed framework in comparison with some alternative lossless compression algorithms. The offered operation sequence is mathematically simple and does not demand essential computing resources.

In prospect it is expected to:

- explore various versions of wavelet-decompositions;
- extend the test set by other AVIRIS, LANDSAT, and SPOT-4 data.

References

1. Infrared Imaging Spectrometer. http://aviris.jpl.nasa.gov/
2. Sagnac Interferometer: Theory & Background. https://www.physics.rugers.edu/grad/506/sagnac-interferometer.pdf
3. Klimesh, M.: Low-complexity lossless compression of hyperspectral imagery via adaptive filtering. Technical report 42-163, Jet Propulsion Laboratory (2005)
4. Pizzolante, R.: Lossless compression of hyperspectral imagery. In: Proceedings of the First International Conference on Data Compression, Communications and Processing, CCP 2011, pp. 157–162 (2011). doi:10.1109/CCP.2011.31
5. Wang, H., Babacan, S.D., Sayood, K.: Lossless hyperspectral-image compression using context-based conditional average. IEEE Trans. Geosci. Remote Sens. **45**(12), 4187–4193 (2007). doi:10.1109/TGRS.2007.906085
6. Magli, E., Olmo, G., Quacchio, E.: Optimized onboard lossless and near-lossless compression of hyperspectral data using CALIC. IEEE Geosci. Remote Sens. Lett. **1**(1), 21–25 (2004). IEEE Press, New York. doi:10.1109/LGRS.2003.822312
7. Mielikainen, J.: Lossless compression of hyperspectral images using lookup tables. IEEE Sig. Process. Lett. **13**(3), 157–160 (2006). IEEE Press, New York. doi:10.1109/LSP.2005.862604
8. Mielikainen, J., Toivanen, P.: Lossless compression of hyperspectral images using a quantized index to lookup tables. Geosci. Remote Sens. Lett. **5**(3), 474–478 (2008). IEEE Press, New York. doi:10.1109/LGRS.2008.917598
9. Tang, X., Pearlman, W.A.: Three-dimensional wavelet-based compression of hyperspectral images. In: Motta, G., Rizzo, F., Storer, J.A. (eds.) Hyperspectral Data Compression, pp. 273–308. Springer, New York (2006)
10. Christophe, E.: Hyperspectral data compression tradeoff. In: Prasad, S., Bruce, L.M., Chanussot, J. (eds.) Optical Remote Sensing, pp. 9–29. Springer, Heidelberg (2011)
11. Chang, C.-I.: Hyperspectral Data Processing: Algorithm Design and Analysis. Wiley, Hoboken (2013)

Author Index

Printed in the United States
By Bookmasters